Developing Science

in the

Primary

Classroom

Wynne Harlen
Sheila Jelly

Oliver & Boyd

Photographs by Andrew Lambert

Oliver & Boyd
Longman House
Burnt Mill
Harlow
Essex CM20 2JE

An Imprint of Longman Group UK Ltd

ISBN 0 05 004258 0
First published 1989
Sixth impression 1992

Set in Linotron Palatino 10 on 12pt.

Produced by Longman Singapore Publishers Pte Ltd
Printed in Singapore

The publisher's policy is to use paper manufactured
from sustainable forests.

Contents

Foreword

It is now national policy in England and Wales that science is taught as a core subject from 5 to 16. There is wide support for this, but unless we want to see science being taught as separate lessons by specialist teachers, it means that all primary teachers must be concerned to include science in their children's work. This book has been written in the belief that all primary teachers not only should teach science but that they can do so with satisfaction and enjoyment.

One reason for this belief is that the main obstacle for many teachers is lack of confidence specifically in relation to science. These teachers do not lack confidence in other areas of their work and experience shows that their skill and confidence can quite rapidly spread to science, once a systematic and conscious start has been made. The word *conscious* is particularly relevant here, since many teachers have already made a start in terms of providing content with science potential, but may not realise it. Once this consciousness appears it can develop into the ability to recognise and pursue science activities even when these are integrated into broader work.

Development is the theme of this book; development in teachers of a progressive understanding of what science is and what is their role in teaching science and through this teacher-development the provision of opportunities for children to develop scientific skills, attitudes and concepts. As for children, development of teachers starts from where they are at present and as for all development there has to be progression through various stages, the earlier ones being a necessary base for the later ones. Thus there is a progression in this book, taking the reader from a focus on particular activities, to the analysis of these activities and suggestions for dealing with the problems arising, to ways of developing activities with other content and, eventually, to looking at the children's progress and planning ahead. In all of this the context is the classroom and the focus the individual teacher, for although the best framework for teacher development may be one where

problems are tackled in groups, it is not possible for all to find this context for their efforts.

The book begins with an attempt to put in to plain and non-threatening words the meaning of science and technology and of *learning* science and technology. These meanings are implicit throughout the rest of the book and indeed provide the basis and rationale for the activities and teaching approaches which are proposed. Chapter 2 asks readers to take stock of what is already being done in their classrooms that can provide potential for science. Some practical investigations for children are then introduced and it is important for readers to try these with their classes, or a group, since later chapters draw on this experience.

The process of reflecting on what the children and the teacher were doing in the suggested investigation is begun in Chapter 3, where checklists are provided for reviewing the children's and teacher's activities. They are followed by ways of interpreting what is found and suggestions for tackling problems which might be revealed.

Chapter 4 is concerned with developing process-based activities from a range of classroom starting points, through a technique of identifying 'skill-scanning questions'. Pros and cons of different ways of incorporating science activities into the rest of the curriculum — as separate lessons, as part of topic work or 'as it arises' — are discussed in the first part of Chapter 5, whilst the later part deals with resources and classroom organisation. Chapter 6 takes a step back from the details of practice to examine the underlying rationale for the kind of work proposed in earlier chapters. It shows that certain kinds of experiences and classroom roles are necessary in order for children to have the opportunity for process-based learning. Chapter 7 sustains the focus on children's learning and development, providing a description which can be used in observing and helping children's progress. Finally, Chapter 8 proposes ways in which planning can be extended to give attention to concepts and content, as well as to process skills.

1. That word 'science'

One thing that has changed in the primary science scene in the last five to ten years is that there is no longer any need to justify the place of science in the primary curriculum. A good case was made out in the late sixties and early seventies through the work of curriculum projects such as Science 5/13, and its value was evident in those classes where it was introduced. The need for more children to have an early opportunity to develop their understanding of the world around them and their ways of exploring it prompted HMIs, in their Primary Survey (1978), to deplore the fact that so few schools had 'effective programmes for the teaching of science'. Then, in 1985, the DES underlined the importance of including science in children's education from the start, urging that: 'All pupils should be properly introduced to science in the primary school.'

The problem now is not one of convincing advisers, heads, teachers and others concerned with children's primary education of the importance of science but of helping with the undoubted difficulties of getting started and keeping going. You may, like many teachers, feel that you ought to be doing some science, or more science, or different science, with your class — and you may even feel guilty about it — but that doesn't make it any easier to get a clear idea of what science *means* and what to do about it.

Possibly the word 'science' has been a source of worry to you. It can conjure up a range of often over-awing images and is not uncommonly associated with some, not always pleasant and successful, experiences in the secondary school. Unfortunately, it is still rather early for there to be many practising teachers who experienced active, enquiry-based science in their own primary education.

Now, as if to make matters worse, the word 'technology' is added to that of science. But if there is any basis for fearing what is involved in bringing science and technology into the primary classroom, it is simply the fear of the unknown. It is the aim of this book to take away the mystery and to show the meaning of science and technology at the primary level. When you see that science in this context is not about abstract theories and principles and that technology in this context is not about 'high technology', with computers or complex machinery, your fears will appear to be unfounded.

It helps straight away to focus not on the definitions of science and of technology but on what we mean by *learning* science and technology.

Learning science

Let's start with learning science. Essentially, this involves children finding out about something through their own actions and making some sense of the result through their own thinking. A child finding out how a caterpillar moves by watching it carefully; another observing a sugar lump breaking up and dissolving in water; the one watching the colours separate out from an ink blot on damp filter paper; all these activities are part of learning science. The *actions* may seem little more than passively observing, but note that there is always some action before there is anything to observe: the caterpillar has to be placed on a chosen surface, the sugar lump put in the water, the filter paper, water and ink put together in a certain way. Observing is also a more mentally active than passive activity, for when we observe, all of us pay attention selectively to some things rather than others and we try to make sense of what we find as we take it in.

Often action is more obviously a part of the finding out, as when the caterpillars are fed in a controlled way to find out if they have food preferences or the blades of a model windmill are trimmed to try to make them turn faster, or floating plasticine boats are loaded to see if some shapes hold more than others before sinking. In these cases the children are physically active, more obviously *doing* than just watching, but we shouldn't run away with the idea that this physical activity is the same as science activity. Children can be busy without their activity being purposeful in terms of learning science; conversely, they can be doing a great deal of learning science through observing, reflecting, discussing and reading — when they are mentally active making sense of things.

As you probably realise, a mixture of physical and mental activity is what is needed. The essence of science activity, however, and usually its starting point, is the encounter between the child and some phenomenon; some face to face interaction of children and things around from which they can learn directly through their own physical and mental activity.

> It may be the clouds in the sky, or the birds in the undergrowth; it may be a bumblebee on the clover, or a spider in a web, the pollen of a flower, or the ripples in a pond. It may be the softness of a fleece, the 'bang!!' in a drum, or the rainbow in a soapfilm. From all around comes the invitation; all around sounds the challenge. The question is there, the answer lies hidden, and the child has the key.
>
> *(Jos Elstgeest, 1985, p.10)*

The child has the key because, as in all learning, no-one can put ready-made ideas and ways of thinking into his or her head. But whether or not children have the opportunity to develop ideas and ways of thinking for themselves depends crucially on their teacher. When children explore the spider's web, the soap bubbles, the ripples on the pond, the ideas they have about these things afterwards will be different from the ideas they had before they had these encounters. They will learn something. But what they learn will depend on many things, particularly on what ideas they had at the start, what they did and how they interpreted what they found. In turn, what they do and what they find will depend on the materials that their teacher provided for them to use, the guidance they received and the encouragement to do such things as think things out, check ideas by going back to the objects, improve their technique for finding out, challenge preconceived ideas. For all these things it is the *teacher* who holds the key.

Learning technology

Let us be clear from the start that what is described these days as technology at the primary level has been included for many years under the label of primary science as advocated by projects such as the Nuffield Junior Science project, Science 5/13 and some later curriculum developments. Science and technology are not the same, but they are so interrelated that they are often closely linked, particularly when we are talking about learning in science and learning in technology.

Science, as already discussed, is concerned with understanding the way things are and why they behave as they do. Technology is concerned with finding practical solutions to problems, especially creating something which meets a human need.

Examples of technology in daily life abound, from building bridges and roads, to making artificial heart valves and producing records and tapes. In the primary classroom examples are creating working models, devising a home-made timer, making a contraption to test the strength of hair or to compare the strength of nutshells. In all these cases the application of scientific ideas is combined with creative design and in most cases an element of aesthetic appreciation is involved, to make products which are pleasing to look at as well as functional.

So when children are designing and making something, answering questions such as 'How can we do this?' they can be said to be involved in learning technology; if they are experimenting and investigating to answer questions such as 'What happens if?' or, 'What's different and what's the same about these?' they are involved in learning science. It isn't difficult to see that answering the 'how to' questions involves applying ideas learned through answering 'what happens' questions. Similarly, when a product has been achieved — the model windmill or timer constructed — it can become the object of investigation and further learning in science. Hence the interpenetration of the two kinds of activity.

This discussion of the distinction between learning science and learning technology shows that we can recognise their differences but without separating them in children's learning. Both are necessary to achieve different aims. You will probably realise that 'primary school technology' is just a label which more correctly identifies some activities which you and your children may have been engaged in for some time under another name.

What is learning in science?

The value of the kinds of activity which have been mentioned and illustrated above can be summarised in broad terms as *helping children to understand the world around them*. But almost all that they do contributes to this understanding and we have to be a little more specific if we are to identify the particular contribution of science and technology activities.

We have described many of the science activities in terms of children 'finding out.' *What* they find out will be such as :

the characteristics of things (living and non-living) around and how these can be used to classify and label things;
how things (again, both living and non-living) behave or work or interact with other things;
what is needed to change something from one position, state or form to another.

These ideas are ones which children can use in solving problems. But the practical solution of a problem is not the only value of tackling it, nor is what is found the only value of the 'finding out' activities. The *process* of doing these things is also a learning outcome, and so children are also learning about:

how to find things out by investigation and observation;
how to test ideas to see if they fit further evidence;
how to apply ideas from one situation to solve a problem in another situation;
how to set about finding practical solutions to problems.

Whilst involved in the processes of finding out and solving problems (and learning how to do these more effectively) children will implicitly be encouraged in willingness:

to find out rather than use preconceived ideas;
to take note of all the evidence in developing and testing their ideas;
to be critical of their own ideas and ways of working;
to realise that they can learn through their own activity and gradually to take on responsibility for their learning.

What we have in these three lists are the *ideas or concepts of science*, the *science process skills* and the *attitudes of science*. Spelling them out in this way may help to show what learning we are aiming for and that this learning is closely related both to common sense and to the aims of other areas of the curriculum. The aims can of course be specified more precisely and we shall need to do this later. If, however, you would like to look at the more detailed list now, turn to Appendix l, page 62.

Learning in and out of school

Children don't only explore, investigate, solve problems and form ideas in school; they do these all the time. They will learn about things in the world around them even if science isn't included in the school curriculum. So why is it necessary? An answer is to be found in some of the 'everyday' ideas which children have which seem rather strange and unscientific: that switching on a light in a darkened room is like switching on your eyes, that snails are slugs which have parked their shells, that things which are small and light float on water regardless of what they are made of, that things which dissolve in water disappear entirely. All of these ideas are reasonable first ideas which children will form from limited observations. It is when these ideas persist and are not modified or replaced by more helpful ones that children encounter difficulties in understanding their surroundings. Science activities at school can help in the way ideas are formed, or existing ideas tested and changed in the light of evidence, so that children are not left with their 'everyday' ideas unchallenged.

Just how this is done cannot be put in a sentence. As with all learning, the starting point is where the learners are, so it is important to be able to recognise this and to provide children with the appropriate challenges and opportunities for advancement. We shall give some help with this in a later chapter: but the starting point for you, as a teacher, is earlier than this and so we consider in the next chapter the position from where you begin.

2. Making a start

Where do we start?

The short answer to this question is 'Start where you are now'. This may seem glib but it really is the key to building confidence in teaching science. Although some classrooms are better endowed than others, every classroom has good potential for science work. As a consequence, all teachers, whatever their experience or anxieties have a well-set stage for science action.

In response to the statement 'Start where you are now', you might ask 'How do we know where we are?' Consider the following response of one head teacher:

> 'We don't do much science. I've one teacher very keen on Nature Study and most of us do some of this work during the year. Some of us attempt some physical science but we are all worried by it. We don't feel confident that we know enough about it and, let's face it, when we add that problem to the difficulty of organising practical work with large classes, it's not surprising that we neglect science.'

This comment is likely to strike a chord. Certainly many teachers share feelings that science:

- is a difficult subject because it appears to require specialist knowledge;
- makes organisational demands that seem daunting, given all the pressures of work in primary schools.

Perhaps you share these views? If so it is important to realise that although these anxieties are real, they can be overcome. Whilst not wishing to minimise the reality of such concerns for some teachers, it is encouraging to know that real potential for science already exists in normal primary practices. For example:

- all classrooms, however cramped, contain a range of everyday materials that are used for various purposes;
- all children handle materials in the course of

their normal work;
- all teachers talk with children about their work.

The two commodities, *materials* and *talk*, are present in every classroom and the answer to 'Where do we start?' becomes one of being prepared to explore and exploit them to engage children in practical investigations and problem-solving.

The DES policy statement *Science 5-16* has stated the value of this activity in these words:

> ... pupils need to grow accustomed from an early age to the scientific processes of observing, measuring, describing, investigating, predicting, experimenting and explaining. Appropriate work can and should begin in infant classes. Pupils should also use their science in technological activities which pose realistic problems to be solved and involve designing and making. Science and technology in the primary school should form, and be experienced as, a continuum. (DES, 1985, para. 24)

For teachers, the practical implications of this are gaining experience of:

- the particular ways in which materials can be explored by children in order to promote scientific activity;
- the kind of classroom talk that can stimulate and structure children's scientific experience.

The key ingredient for implementation is the willingness of teachers to 'have a go'. What follows is selected to help you gain confidence to do just that.

How do we start?

We will start with a problem:
which fabric is best for keeping us dry?
— a good problem for scientific investigation by

children of all ages. There's a variety of ways in which it might be introduced:

> *semi-spontaneously*, where preference is for responding to the immediacy of topical interests. For example, a sudden interest-capturing downpour in a spell of dry weather when many children have come to school without rain-day clothes; or, perhaps, when someone has come to school resplendent in a new raincoat that invites comment;

> *pre-planned*, as an activity within a wider topic: weather, or clothes and clothing, for example;

> *deliberately*, in a 'today we are going to be scientists' style; an approach that works well with juniors;

> *indirectly*, as a challenge for some children to respond to in non-class time if you prefer not to undertake the work in class time. The challenge might be presented as part of a display, on, say materials, or it can be effective, standing in its own right, as a new classroom event with novelty appeal.

No matter how the problem is introduced, the activity that follows will need planning in terms of *resources, a teaching sequence* and *class organisation*.

Resources

Essentially children will require:

1. *Some fabrics*
   ideally, acquire some old clothes from a jumble sale (or ask children to bring in old clothes or scraps of cloth);
   more conveniently, raid the school fabric box.
   Whatever the source, it's useful to limit the range of fabrics available during the activity. Three or four different kinds are usually all that children new to this work can cope with. It's not necessary to name the fabrics (though you may feel happier to select those known to you should the children ask).
   Do some private investigation to decide the best selection of fabrics. Try dropping water on them and include in your selection one that

water goes through fairly readily and one, e.g. felt, on which a drop of water forms a ball on the surface and is only slowly absorbed. It's a good idea to avoid fabrics that children know to be waterproof, otherwise any investigation may seem pointless to them and they are likely to respond in a 'this one's best — so what?' manner. The approach is along the lines ' . . . if *these* are all we have, which would keep us driest?'

2. *Some means of simulating rain*
   courageously, use mini-watering cans;
   less dramatically, use any container that the children can manipulate competently for pouring. With potential spillage in mind it's sensible to keep the containers as small as is conveniently possible (old film-spool containers, for instance).

3. *Some means of seeing what happens when water falls on the fabrics*
   it is important to respond to, and let the children try out, any ideas they have (such as placing the fabric on their hand and feeling if water comes through);
   but, ·.
   should ideas not be forthcoming, then 'stage-manage' the situation. For example, use the approach 'perhaps these would help' and simply supply suitable materials, or, more directly, 'should we try this?' and, with them, place fabric over a see-through container (plastic party glasses are good) and hold it in place with an elastic band.

A teaching sequence

Having sorted out the kind of resources needed it is now necessary to have a teaching sequence in mind before sorting out the thorny problem of class organisation.
   The work will fall into three phases:

> setting up the problem;
> preliminary exploration;
> investigating.

Setting up the problem

This is a short verbal phase with three aspects of development.

1. *Make sure that the problem has meaning for the children in terms of their own experience.* Some teachers favour an initial focus on the clothes the children are wearing using a speculative approach of the kind:
'If we were out without coats and it poured with rain, whose jumper/sweater would be best for keeping you dry?'
The question will promote useful talk.

2. *Develop the meaning phase with a focus on the actual materials to be tested* via a question of the kind 'Which of *these* would be best?' Collect opinions.

3. *Establish that practical action is needed to put opinions to the test.* Most likely there will be a range of opinions in a group and the situation can be structured to favour activity via 'I wonder who's right? What could we do to find out?' Should there already be unanimous agreement on which would be best then suggest, 'What could we do to find out?'

Preliminary exploration

In this phase children will be exploring what happens when water falls on the different fabrics and it is important that they do this in their own terms. They are likely to be fascinated by specific events, such as the 'balling' of water on felt and in responding to such particular happenings they may well lose sight of the original problem. What happens may lead some to explore tangentially, dropping water on other things; some may carry out repetitive drop-making, exploring big drops and little drops. Few, if any, are likely to work systematically in an attack on the problem. This is usual for children inexperienced in scientific investigation and is, in fact, a diagnostic indication of the level of investigative skills. For this reason the preliminary exploration should not be rushed, nor should it be over-structured by teachers anxious to move quickly towards 'the right answer'.

For some children this exploratory phase may well be sufficient as an early scientific experience, but it is useful to help them draw together what they have done and encourage comparison between the fabrics by asking 'What happens when . . .?' They might classify the fabrics as good/not good for keeping water out, or put them in rank order of best→worst.

Once comparisons have been made the scene is set for the next phase, of investigating.

Investigating

The key idea here is that of the 'fair test'. When children say that one fabric is better than another for keeping rain out, have they been fair to the fabrics in their testing? What might have made their comparison unfair? It is important to find out the children's ideas on the subject. Probably someone will identify that to be fair they need to use the same amount of water on each fabric, though this will not necessarily happen. Lack of response is another indication of the level of the children's investigative skills. It might also be that the whole idea of fairness in this situation is so new to them that they are not sure what response is wanted, so it's a good idea to probe with questions of the kind 'Does it matter how much water we used?'

Work towards suggestions from the children of how to treat the fabrics in a fair way and then let them carry out their test, or repeat it if necessary, using their ideas.

Look for opportunities to focus the work towards measurement, for example, 'How much better is this fabric than that for keeping out the rain?' But beware of moving too quickly in this direction. Most children will not be ready for precision work until they have had a considerable amount of experience of general comparative testing.

Class organisation

Let's now consider ways of organising the work. There are many possibilities, but ones suggested for making a start fall into three kinds:

(a) starting with one, largish, group only;
(b) starting with the whole class, which is then divided into smallish groups;
(c) starting in out-of-class time.

One large group

Starting with a group of about six children can be difficult in some situations, for the rest of the class need to be doing work that is not too dependent on the teacher. With a class unused to practical group activity some teachers have:

- set the other children writing/reading tasks and handled their natural interest in the work of the practical group with the response 'your turn will come';
- adopted a similar approach but with the rest of the class doing art and craft work.

The whole class, divided into smallish groups

Two organisational structures used by teachers, which start with the whole class but find different ways of working with small groups are summarised below:

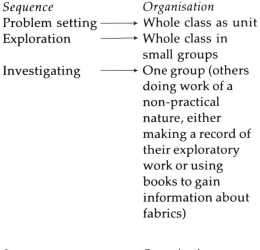

Sequence	Organisation
Problem setting ⟶	Whole class as unit
Exploration ⟶	Whole class in small groups
Investigating ⟶	One group (others doing work of a non-practical nature, either making a record of their exploratory work or using books to gain information about fabrics)

Sequence	Organisation
Problem setting ⟶	Whole class as unit
Exploration ⎱ ⟶	⎰ Whole class in
Experimenting ⎰	⎱ groups of about six

The second example not only makes heavy resource demands, but it can be difficult to give detailed attention to any one group. Nevertheless, there are also advantages. If the problem has been well set up, the children do not need attention all of the time — indeed it is a positive advantage for them to be left alone to do their own thinking whilst the teacher is busy elsewhere — and the teacher is having only to deal with one type of activity at a time, and not continually jumping from one to another.

Out-of-class time

Far easier, if possible within a school, is making a start without having to cope with organisational pressures, by operating out of class time. Many teachers have gained confidence in the approach by undertaking the work with children in a science or general interest club; this has the obvious advantage of allowing ideas to be tried out with well-motivated children. Later, as confidence grows, the work is drawn into class time.

Unfortunately there are no 'golden rules' and no best patterns for organising the work. We all have to find our preferred working patterns, accepting that the more opportunity we can make for constructively interacting with the children as they work the more productive their work is likely to be.

Summary

In whatever way the work is organised, the guidelines in Fig. 2.1 summarise steps in making a start. In particular, the key questions that help structure children's activity are likely to prove useful.

Be warned

We hope you will enjoy yourself 'having a go'. Don't be discouraged if things don't go as you expect and hope for. On page 14 are some often-encountered problems with some suggestions of reason and what to do in response to them.

Going further

After you have 'had a go' with the fabrics, you might like to 'keep going'. Here are two more problems which will extend the children's experience.

- Which is best for keeping the wind out? (Developed through an interest in fabrics.)

- Which kind of paper is best for covering a book? (A more complex problem that might be developed through an interest in testing.)

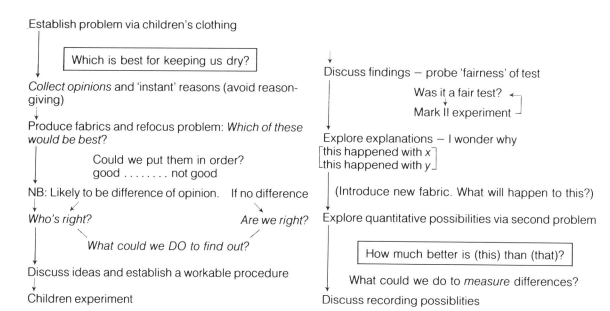

Fig. 2.1 Summary sequence

Problem and possible reasons

Ideas are not forthcoming from the children. They may be uncertain of what is expected of them in a new style of working.

The children find it difficult to sustain interest. Sometimes the desire of teachers to keep the children moving on through the task may take the children out of one phase before they've had sufficient time to gain confidence in working with materials.

The children have difficulty with the idea of a fair test. This is an indication that they need a lot more activity with materials.

The children have difficulty working cooperatively. Even children well used to group working in other areas of the curriculum are likely to have initial problems when they handle and share materials in an investigative manner.

The children ask questions you cannot answer. This happens to even the most experienced and knowledgeable teacher. It can never be avoided as long as children are inquisitive, as they should be, about the things they find around them.

Response

Show that you value their ideas, whatever these are. Be patient, give more experience of exploring materials and sharing ideas.

Give more time for the free exploration of materials. Don't be anxious to turn this into systematic activity. Stop when interest flags and return to the materials later.

Let them work with the materials in their own way, and probe, in discussion, what they think they have found.

Give gentle reminders. Give reasons for the need to cooperate in terms of the scientific activity ('scientists find out more by working with each other' type of approach).

On pages 64-5 you will find some approaches to this problem. Take heart in the knowledge that children's questions tend to change as they gain more experience in investigative work, and learn from your example, from emphasis on information to ones that can be answered by enquiry.

3. How are we doing?

At this point, having 'had a go' at the fabric activities with some or all of your class, it is a good idea to take a little time to sit back and reflect on what you and the children were doing. This is a good thing to do in any situation, for if we are always close to the action and solving moment-to-moment problems of keeping things going, there is a tendency to lose sight of longer-term objectives. It is especially important when trying something new, however, to ask ourselves the question 'Am I on the right lines?' Attempting to answer this question should help to identify, at an early stage, whether or not changes in approach and organisation are needed.

Reviewing the children's activities

We look first at what the children were doing. Think back over the fabric activities and ask yourself the questions in the checklists which follow. Note that the questions are expressed in general terms so that they can be used in relation to other activities as well as to the fabric ones, and so the more general word 'materials' is used to include a range of objects and articles that children might investigate. (It is unfortunate that fabrics are loosely called materials!)

In the Preliminary exploration phase:

were the children handling materials themselves?

did they observe the materials closely — perhaps smelling, feeling, listening, as well as looking?

did they try things 'to see what happens'?

did they group some things together according to what they had found out?

did they use their observations to put things in an order?

did they discuss with each other what they were doing?

did they compare what they found with what others found?

did they discuss their work with you?

did they ask questions or puzzle over anything they found?

were they kept busy and absorbed by what they were doing?

did they base their statements on evidence rather than preconceived ideas?

In the Investigating phase:

did the children have a clear idea of what they were trying to find out or compare?

did they devise a test which was relevant to what they wanted to find out?

did they consider how to carry out the test in a fair manner?

did they actually control variables to make the test fair when they carried it out?

did they use instruments to aid their observation or for taking measurements?

did they make a record of what they did or of their results?

did they discuss their results with each other?

did they look for evidence to support their statements?

were any conclusions reached consistent with their results?

did they obtain and use quantitative results (not just 'more than' but 'how much more than')?

did they check or repeat parts of their investigation?

did they try or discuss different approaches to a problem?

did they criticise what they had done and identify ways of improving their investigation?

These questions are expressed in a way which make it obvious that the best answer is 'yes' to all of them. They indicate some of the things that it is suggested children ought to be doing in science activities. There are good reasons for suggesting these things, for they give children opportunity to develop the ideas, skills and attitudes which were mentioned briefly in Chapter 1 (page 7). Later we will further discuss the justification for the suggestions implied in the lists. For the moment we turn to the implications of the answers to the questions.

There is no problem where the answer is a clear 'yes'. But where there is doubt, or an honest answer is 'no', then some further reflection is necessary. 'No' answers to the 'Preliminary exploration' list are particularly worth studying, for they might indicate that the children were not really engaged in fruitful interaction with the materials. If there are more than one or two 'no' answers it is necessary to recall what the children *were* doing instead of the things in the list. Were they:

- ignoring materials given?
- disputing possession of the materials?
- looking only superficially at the materials?
- spending most of their time writing or drawing?
- just watching others and not joining in?
- using materials in unintended ways?
- moving around the classroom fetching and carrying materials?
- talking about things not related to the task in hand?
- doing anything else?

Answering 'yes' to some of these questions could indicate that more attention might need to be given to how the problem was set up. Some possible sources of difficulties include:

- perhaps not enough materials were provided;
- perhaps the children's interest in the particular materials had not been sufficiently awakened;
- perhaps they did not understand what they were supposed to be doing;
- perhaps they needed more help to realise that they can find things out for themselves through their own actions;
- perhaps they view school work as being essentially a matter of writing.

Suggestions given on page 17 may help where these difficulties apply.

In the 'Investigating' phase, 'no' answers to several questions in the checklist are perhaps to be expected, certainly at the start. Some items on the list (e.g. checking and repeating parts of an investigation) refer to things that it is known children tend not to do when left to themselves. Others (e.g. using measuring instruments and collecting quantitative results) are more relevant to certain problems than to others where qualitative results are entirely adequate. Indeed, all the items in this list indicate more advanced levels of skills and attitudes than is the case for the 'Preliminary exploration' list.

Although we don't expect a full set of 'yes' answers at the start, we are, of course, concerned with what can be done to work towards advancing children's investigative abilities. This is something we will take up later, in Chapter 7, where we discuss children's progress in particular ideas, skills and attitudes and consider ways of helping this progress. For the moment, however, our focus is at a more general level, remembering that we want children to:

- think for themselves about their investigations;
- do the investigations for themselves;
- work out for themselves what their results mean;
- review and improve for themselves what they have done.

Chapter 2 mentioned *materials* and *talk* as key starting points for science. We now add two more key factors in giving opportunity for advancing children's learning: *time* and *discussion* (a particular form of talk in which the participants respond to each other on an equal footing).

The time needed for children to do things for themselves is always much greater than if other people do them for the children. This applies to thinking as much as to tying shoe laces. But, as with tying shoe laces, children will never learn to think for themselves if someone else always does the thinking for them. So we must expect that deciding what to do, working it out in practice and then doing it will take children longer — perhaps two or three times longer — than following instructions which tell them what to do. But in this time they will be learning much more science in one activity which is thought through, than if they filled the same time doing several activities where they didn't have to think for themselves.

Not only do we need to provide the time, but to make sure it is well spent — for *thinking* as well as *doing*. This is where *discussion* is so necessary, for it provides both:

a vehicle for teachers to encourage children to think aloud, to do their own thinking (and to realise that they can indeed do this) and not depend on instructions from others;

a channel for children to reflect critically on their work, both in terms of how well results are backed up by evidence and how procedures could have been improved.

If discussion is to serve these functions it is useful to remember that it is:

an interchange in which all involved can put forward ideas, ask and answer questions, on an equal basis;

helped by teachers asking open questions and ones expressing interest in children's ideas ('What do *you* think about . . .?').

Reviewing the teacher's activities

In responding to reflection on what the children were or were not doing, we have already indicated actions that you might consider taking. Now we look more directly at the teacher's role and list some of the aspects which initially present difficulties to many teachers. If you share some of these then some of the suggestions for tackling them may be worth trying.

Common areas of difficulty in the 'Setting up' phase

Engaging children's interest in the problem as a real one

Suggestions: Setting the scene is important here, as is timing. For some topics it is useful to mount a display of relevant objects in the classroom. Encourage the children to touch

and talk about the objects and add some of their own. Add something yourself each day prior to starting a class discussion. In the case of the waterproof fabrics, the weather will probably oblige with a convenient downpour. Start the discussion by referring to this and to the appearance of anyone coming into the school showing signs of having been in the rain.

Transferring the children's interest from the general problem to one which can be tackled with the materials provided

Suggestions: Spend more time on general discussion before 'homing in' on the problem. With the raincoats, for example, have a parade of rainwear; divide the articles into groups according to features noticed by the children; regroup by different criteria; choose two very similar coats and list the differences; choose two very different ones and list the similarities. Which is most waterproof? (We don't know.) Could we find out? (Better not to experiment on the coats.) Let's see if we can decide this question for these pieces of fabric . . .

Encouraging children to 'have a go' rather than ask for answers

Suggestion: Some children need an invitation to start interacting with the materials, before a specific problem is posed. So the question 'Can you find out which is best . . .?' or 'How can you test . . .?' may be too far ahead of these children. Instead, try comparison questions such as 'In what ways are they the same? In what ways do they differ?' and 'What happens if . . .?' questions first.

Common areas of difficulty in the 'Preliminary exploration' phase

Responding to unexpected things that children do

Suggestion: If we want children to use their own ideas we shouldn't be surprised, but pleased, when they do something unexpected. However, there's no doubt that it can be disconcerting, for instance, to find the children sewing the fabrics instead of testing them with water! Unless something dangerous is likely to happen, though, stop yourself 'jumping in' straight away. Observe carefully from a slight distance so that you know just what is happening and then ask the children to explain to you their view of what they are doing and what they are hoping to achieve through it. It is likely that they have a good reason — good in their minds — for what they are doing and, if not, in telling you about it they will probably realise their mistaken reasoning. (The children in question thought that if the fabrics they were given were too tough to sew or would tear at the seams, they'd be no good for a raincoat anyway!)

● *Answering the children's questions*

Suggestion: Handling questions depends on the type of question, so it is difficult to sum up at a general level in a few words. The discussion in Appendix 2 should help.

● *Supplying additional materials requested by the children for their explorations*

Suggestion: With experience it becomes easier to anticipate children's need for containers, sticks, cardboard, pieces of tubing, etc. Many teachers find it helpful, as well as being useful to the children, to ask them to plan ahead what they will need and to write it down during one session of science activities in preparation for the next one.

● *Judging when to bring this phase of the activity to a close*

Suggestion: Let the children help you, indirectly, in judging the point for rounding off an activity, or a phase of it. If the activity appears to be waning, or losing focus, give the children a few minutes warning to prepare an oral or written account for you of what they have done. This account will help you to judge whether they have already done as much as you could expect or whether they have not really come to grips with the problem. You can then take the appropriate action.

● *Ensuring equal participation of all children in a group*

Suggestion: Where there is a problem of ensuring equal participation in the activity among the members of a group, it is sometimes necessary to discuss the problem with them. Help them to agree to defining tasks within the activity and to a fair system of taking turns at these tasks. If the problem persists it may be necessary to experiment with changes in group membership.

Common areas of difficulty in the 'Investigating' phase

These might include all those in the 'Preliminary exploration' phase plus some others. Many of the points made previously will apply in this phase. The additional ones arise because we are concerned with rounding off the activity as well as with keeping it going.

● *Encouraging children to make some appropriate record of their work*

Suggestion: It is helpful to think of any record the children make of their work in the context of their whole *communication* about the activity. Discussion, notes, drawings, models, paintings and a display of work are all parts of this communication and each has a different role to play. There is nothing special about written work and it should not be asked for as a ritual. Thus, at times, particularly in the early stages of 'getting going' in science, it may not be necessary to make a record. Instead there should be plenty of talk to help children review and reflect upon what they have done.

That said, however, words do have a particularly efficient function in communication and we want children to become used to and proficient in using them. Among written forms of communication there are both personal notes and public reports to consider. Personal notes are necessary to assist in planning, to record things that may be forgotten and for taking down results. A long-term aim is to encourage children to use notebooks to jot down, for themselves, notes, drawings, measurements, which will help their investigation. These are not the kind of record that people usually mean when they talk about recording work, but they are a very important part of the activity.

It is the public report, the kind that emerges only at the end of an activity, which is usually

of more concern to teachers. It is much less of a problem, however, when it is seen by both teacher and children to be *relevant* and *appropriate*. Relevance comes from having some function, which is usually that of telling others about what has been done. So it is necessary for the record to be seen to be used for that purpose; not just pinned on the wall (if it is on paper) but talked about by those who produced it; not an isolated artefact but one which the makers explain, demonstrate and show how they solved practical problems in its construction.

Appropriateness has a double significance in this context:

- what is appropriate for the children at their stage in active learning;
- what is the best medium for the particular information to be communicated.

It may be a chart, a graph, a model, etc., and it may only need a few words of explanation or invitation to a potential audience. As children's work becomes more sophisticated, they themselves will find that it needs more words of explanation and then some structure in reporting will be appropriate, but this should not be introduced prematurely.

Bringing the activity to a conclusion

Suggestion: The making and communication of a record is one part of bringing the activity to a conclusion. It is best seen in the context of exchanging reports from different groups and giving all children the chance to see the work of the class brought together and to comment on it. Children are always interested in what other groups have been doing, especially if there is a linking theme but they have not been doing the same as each other. It is useful to establish that a whole-class discussion will generally be held at the end of a topic, and that groups will be expected to show and describe their work and findings. Preparing for this brings relevance to the children's reporting and makes them consider an appropriate form for the communication. It also ensures that work is rounded off by the group, not stopped in mid-investigation, and it makes them review for themselves what they have done in preparation for describing it to others. Sometimes the combined reports and products can be put together in a display in some public area of the school for other classes, parents and visitors to see.

4. Developing the work: getting ideas

In Chapter 2 we described, in some detail, one schedule for making a start on investigational work and in Chapter 3 we analysed the experience. Naturally you will make judgements about the relevance of the work to your own class situation and may share this teacher's view:

> 'The children enjoyed working with fabrics. I found it interesting and the checklists helped me work out ideas about the kind of science I should be teaching. But when I think about how this experience helps me organise science for all my children over a long period of time, the mind boggles!'

In this and next chapter we acknowledge the 'mind boggling' aspect of planning science for a whole class within a whole curriculum, for a whole year, and offer practical suggestions for approaching the task.

Planning science work

The DES policy document *Science 5-16* (DES, 1985) requires that all primary teachers incorporate science into their programmes. That's fine for the children and a reasonable requirement of teachers, but it has its problems.

- How, in an unfamiliar subject, do you get ideas for developing children's science work?
- How do you fit science into what seems an already overcrowded curriculum?
- How do you cope with resourcing and organisational demands?

These are real and appropriate concerns for those who lack experience of teaching science in an investigative manner.

Here we will deal with the first of these questions; the others will be taken up in Chapter 5.

Getting ideas for children's activities

This aspect of planning may seem formidable but it becomes relatively easy if we adopt the following approach. Essentially what is needed is experience and practice in generating ideas for science work by:

1. identifying the science potential in everyday things and situations;
2. converting this potential into *action questions* that can shape children's investigations.

What, then, is involved?

Seeing the potential

Let's take as an example possibilities for work on *buildings*. Much general potential for science can be identified if we scan the topic in terms of its opportunities for skill development (a skills-scan), bearing in mind that these skills relate to obtaining and organising evidence (discussed later in Chapter 6) and require the children to have direct contact with materials.

Hence the first planning decision is a fairly obvious one, namely, *what* can the children investigate? Possibilities are:

    building materials;
    actual buildings;
    building-site activities.

Given all or some of these resources, a skills-scan involves asking a number of questions to identify *general possibilities*. Table 4.1 provides an example of what might emerge.

Notice that:

- for scanning purposes a short list of skills has been used. This is useful for first-stage planning; other opportunities relating to the full list of skills (see Appendix 1) can be accommodated at a later stage;

- the possibilities are speculative. This is because they are not yet shaped sufficiently to indicate particular investigations children could undertake. For this to happen they need 'converting' to specific challenges.

Producing action questions from general possibilities

The ability to convert general possibilities into action questions for children to tackle is probably the most important skill in planning science work. Like other skills it develops well with practice and it becomes a comparatively easy task if approached systematically.

We will develop the idea by considering particular possibilities for investigating a school building, since this is a resource available to every teacher. Systematic thinking to identify action questions will run along these lines.

'If *X* is what children encounter and *Y* is the general skill possibility I have in mind, then what questions will initiate activities so that the children:

have a problem to investigate?
can tackle the problem in ways that foster that skill development?'

You may find the use of *X* and *Y* somewhat off-putting, but let's see how the approach can work out in practice.

Possibilities for observation

If *X* is the school building and *Y* the observation possibilities from Table 4.1, namely noticing detail and pattern making comparisons, then action questions abound.

- How many different kinds of materials can they see? (Note that this question is preferable to one that asks children to name the materials present.)
- Which materials can be seen in many places? Which in only a few? Are any found in one place only?
- How many different shapes? Which is the commonest? Which is the least common?
- How many different parts?
- Which things are solid? Which hollow? Which let things through? Which keep things out?
  Which join things? Which separate things?
- Are all similar parts exactly the same? Do different parts have things in common?

Of course *X* could be part of a building, say a wall, rather than the entire structure and in this case action questions that encourage observation include:

- Are all bricks (stones) in the wall exactly the same?
- How many differences are there between two bricks of the same kind? How many similarities?
- How many bricks does one brick touch?
- How many different things are there on the wall/in the wall?
- Is there any pattern in their distribution?

The emphasis is on producing questions that will get children looking closely so that they see things not noticed before and experience things which help them to build up *ideas* of similarity and difference, of relationships between whole things and their parts and of causal rather than chance distribution in person-made structures.

None of the questions identified requires of teachers in-depth scientific knowledge and most are of the kind that would be asked by any teacher wishing to engage children in first-hand observation. Indeed it is difficult to claim that *these* questions are particularly *scientific* — they could equally well lead to activities in other areas of the curriculum. But their relevance across the curriculum does not diminish their value as precursors to activities of a more scientific nature. Activities involving testing, design-and-make work and predicting are at the heart of the science/technology curriculum. They tend to be lacking in many children's experience because teachers are uncertain of how they can be developed. The approach we are adopting can, by starting from observation, stimulate many other possibilities for scientific activity.

Skill-scanning questions	General Possibilities
What might they *observe?*	Detail and pattern in particular building materials and in buildings and in their parts? Comparisons between different building materials and different building constructions?
What might they *investigate* (test)?	Properties of building materials? Of building tools? and machines?
What might they *design and make?*	Build a simulated house? Make machines for lifting, carrying, etc.?
What might they *predict?*	Changes in building materials? How long will this take? (For example, concrete to set, etc.)

Table 4.1

Possibilities for investigating (testing)

Table 4.1 notes as general possibilities under this heading testing the 'properties of building materials'. There is tremendous potential here, but how to convert it into action questions?

We need a framework for thinking, along the lines of the examples below.

● What are the significant properties/features of the materials?

One significant property of a wall is that it is strong. Children will be aware of this, at least implicitly. They can be encouraged to observe different patterns used in the construction of walls. Suitable action questions would then be:

Which pattern is strongest?
Is this pattern stronger than that?

● What situations will encourage children's awareness of these properties/features so that they can appreciate variables and attempt to control them?

A significant feature of most walls is that the bricks or stones are joined together by something (mortar). Children can observe this directly. They can be encouraged to find ways of joining bricks/stones using mortar. Some action questions would be:

Which method makes the strongest join?
Which mix sets first?

● What forms of question will focus their work towards handling variables?

By applying the approach with X as the school building and Y as investigating (testing), a variety of action questions emerge. They can be grouped into possible general forms:

Which . . . is best for . . .?
e.g. Which method is best for protecting woodwork?
　　　Which product is best for cleaning windows?
　　　Which mix makes the strongest concrete?
Is it true that . . .?
e.g. reinforced concrete is stronger than non-reinforced concrete?
　　　smoke rises further from tall chimneys than from short ones?
Which . . . is/does . . .?
e.g Which door is easiest to open?
　　　Which window lets in most light?
　　　Which kind of brick absorbs most water?
　　　Which concrete mix sets faster?

With practice, questions such as these form quickly, whatever the nature of X.

Possibilities for designing and making

Let's turn now to 'design-and-make' work. General possibilities, as recorded in Table 4.1, emerge by considering the particular constructions and 'things that work' associated with a topic, but further thinking is needed to convert these into action questions.

  What experience will children need to appreciate the structure/function of the real thing?

For children to appreciate the construction of a building they need to have observed its component parts and to have some appreciation of the sequence of events in its development. This latter can be achieved ideally through watching building-site activities; more conveniently by learning about events from information books.

  What might they do/use to simulate its structure/function?

The sequence
building plot → foundation → walls, doors,
      window spaces → roof
could be simulated on a small scale, say a seed-tray-sized plot. Children could make concrete foundations for walls, use lolly sticks for roof trusses and hang tiles made from thin plastic.

  What are the key questions that will make their work purposeful and not just making for making's sake?

Such questions might be:
  What shall we use for making ...?
  (teacher's ideas are long-stop possibilities)
  How can we make sure it is ...? (as in the real thing)
  e.g. that a wall is vertical?
      that a roof has the required slope?
      that the concrete has the required consistency?

Such questions are appropriate also for simulating things that work. For instance, a machine that will lift a bucket from floor to roof height which, in a simple form, could be made from a cotton reel, string and a yoghurt pot. Its making might be stimulated by seeing the real thing in action or an illustration in a book.

For this, as with all 'working things' children make, a further key question is:
  How could we make it work better?
  e.g. lift more quickly?
      lift more steadily?
      lift with less effort?

Again, ideas will come readily with practice.

  Further possibilities for making working

things relating to buildings include:
  What shall we use for making ...?
  e.g. a spirit level?
      scaffolding?
      a crane?
      a concrete mixer?

These kinds of activity need more science knowledge and more resources than observational activities. But experience shows that, almost inevitably, the necessary ideas and most of the resources will come from the children, if their teacher is willing to 'have a go'.

Possibilities for predicting

Predicting activities are ones which involve children going beyond their present experiences and observations in some way. They must be firmly rooted in these experiences and enable children to use and test out ideas they have about what they have seen or done.

They take the children from exploring *what is* to exploring *what might be*. For example:

  Would a wall twice as thick be twice as strong?
  (Based on the idea, supported by observation, that thicker walls are stronger, and leading to a prediction that a wall twice as thick would be twice as strong.)
  Will the mortar set more quickly if we use warm water?
  (Based on everday experience that heat often speeds up changes — as in cooking — and leading to a prediction to be tested that the warm water will shorten the setting time.)
  Will it be easier to lift the bucket if we use a larger cotton reel for a pulley?
  (Based on an idea about how the pulley helps and leading to a prediction to be tested that the larger the pulley the more it helps).

It doesn't matter that some predictions will be disproved (as in two of the examples above); in fact it is most important to encourage children to test out *all* their predictions, however well founded. As we shall discuss further in Chapter 6, this is essential to the development of their ideas. It is necessary, however, to help them distinguish between a guess, for which they have no basis in experience, and a prediction, which is using their experience.

So, when children make a prediction, or ask a question which implies a prediction, it is a good idea to ask them 'Now why do you think that might happen?' or 'What makes you want to try that?' By doing so not only will you learn a lot about their thinking, but it will help the children to sort out their ideas to try to explain them. But don't make it an inquisition. If the children can't explain, then let them carry straight on to test their idea in practice.

As the examples in Table 4.1 show, some prediction activities have a time dimension. With this in mind action questions related to, say, observing walls, will be of the kind:

Which things are there all the time?
Which are there some of the time?
Will it look exactly the same tomorrow/next week/next term?

Children will need to think about the ques-tions, perhaps recording what they expect to happen, and later to check their ideas against actual events.

Ideas into practice

It should be apparent from the preceding analyses that there is no shortage of ideas for science activities if they are sought systematically. Indeed teachers who regularly use the approach described above find this aspect of planning fairly easy; their major task becomes one of selecting, from the wide range of possibilities, a suitable starting point and anticipating its likely development so that the work overall has a reasonable skill coverage.

Figure 4.1 shows one teacher's forecast for developing a topic on walls with 5-6 year olds. Figure 4.2 is a record showing the sequence of events that actually took place.

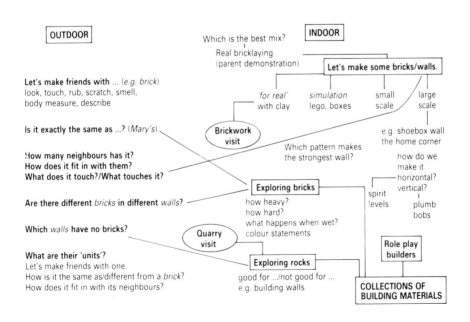

Fig. 4.1   Exploring walls: forecast

Starting point:

Making friends with a brick in the school wall

Rubbings, counting its neighbours
Observing detail
Comparing with a stone in another wall
Finding 'different kinds of' walls
Collecting observations of things in and on walls
Comparing own brick with another 'kind of' brick

↓

A brick collection

Sorting and ordering
Comparing 'ordinary' and sea-worn bricks of the same kind (locating Whitehaven – the name on the bricks)
Collecting 'what will happen if . . .' questions
Collecting 'I wonder why . . .' questions

↓

Making model walls

Free building with big lego
Task building of different bonding patterns
Making spirit levels and plumblines to check horizontal and vertical
Which pattern makes the strongest wall?
Inventing a 'wall-basher'. Discussing fairness of test and modifying

↓

Making concrete bricks

Examining ready-mix concrete; separating and describing parts
Guestimating how many measures to fill the mould
Exploring mixing to produce a 'nearly solid' mix
Predicting when the concrete will set

↓

Making a brick wall

Examining own-made bricks. Noticing differences
Discussing reasons. Comparing own brick with teacher-made bricks
Examining ready-mix mortar. Noticing differences in sand and cement components; relating to ready-mix concrete
Mixing mortar. Using simulated trowels to lay home-made bricks in a stretcher-bond pattern

↓

Relating walls to other parts of a building

'Making friends' with doors, windows, floors, roofs
Collecting observations of similarity and difference in each
Collecting ideas for experiments concerning functions of parts
Sequencing events in building construction via picture cards
House building on a seed-box sized 'plot'. Plans
'Digging' and laying concrete foundations. Wall building

↓

Relating bricks to other building materials

Exploring a collection of building materials
Grouping and sequencing concerning use, properties and location
Collecting 'let's find out . . .' questions
Junk modelling of a building to simulate real locations and functions

↓

Sharing experience and broadening concept of walls

Class forum with questions to panel of 'experts' on walls of buildings and building materials
Collecting and discussing ideas for further investigations of walls via picture situation cards

Fig. 4.2   Exploring walls: record of work

A few significant points to notice in Figs. 4.1 and 4.2 are:

- the use of action questions in the forecast to indicate possibilities;
- the way in which possible activities are linked but not rigidly sequenced;
- that although both forecast and record have the same starting point, the actual development was sequenced to respond to children's particular interests as the work progressed.

(You might also like to analyse the record and decide whether or not it had a reasonable skill-coverage.)

Going further

If the planning approach we have described appeals to you, why not try putting it into practice?

Take the general topic *food* and

1. identify its general possibilities (as in Table 4.1);
2. convert these possibilities to action questions;
3. construct a forecast scheme of work for your class.

5. Developing the work: curriculum and classroom organisation

The planning approach suggested in the previous chapter helps teachers find ideas for children's activities. Of that there is no doubt, but it is equally certain that ideas alone will not get science firmly established in a classroom. However exciting the possibilities for particular activities the work will not develop strongly unless we consider the problems of:

- how to fit science into the existing curriculum?
- what to do about resources and organisation to ensure the work runs smoothly?

These matters are considered in this chapter.

Developing science in the class curriculum

In general, primary teachers have three main ways of giving science curriculum time:

- introducing it 'as and when' an appropriate occasion arises;
- setting aside fixed time for science lessons;
- integrating science with other curriculum areas in topic work.

Each pattern has advantages and disadvantages, which are summarised in Table 5.1. Our concern is not to debate which pattern is best, because as a long-term aim we will want to establish a flexible organisation to accommodate them all. Nor should we become too involved in discussion about how much time science should receive. It's more important to concentrate on the kind of scientific experience children should have and adjust situations accordingly than to start with a particular time allocation that has to be distributed in some manner. The question of science time is not so much about quantity as about the quality of what goes on in whatever time is available.

Science in the curriculum	Main advantage	Disadvantages
'As and when'	High motivation of children	Chancy. Development of situation can be hampered by lack of immediate resources
Separate lessons	Subject receives regular attention	Time restrictions can inhibit investigative work. Fixed time-slot can stifle spontaneity
As part of wider topic	Children's learning can be enriched and made more meaningful through links with other subject activities	Science can 'get lost' in a general topic and work is sometimes superficial. Links with other subjects can be spurious

Table 5.1

Science 'as and when'

Science based on spontaneous happenings (excellent starting points) often remains underdeveloped because it's difficult to organise appropriate resources for immediate response. Hence, over-reliance on the spontaneous can mean that children have little opportunity to undertake sustained investigations and consequently they receive an impoverished diet of scientific experience. One sensible strategy for developing the 'as and when' approach is to identify and anticipate a number of 'guaranteed happenings' and pre-plan for more sustained development. For example:

- in rural and suburban settings there will be many guaranteed seasonal happenings, such as a conker craze;
- in urban settings likely local events involving road works or traffic activity make ideal starters;
- in all schools there will be predictable happenings such as a new baby in someone's family or preparation for Christmas and other festivals.

Whatever the starting point we can pre-plan for its development by using the 'potential →

action question' approach, which will identify the resources needed.

Two such plans produced by 'as and when' teachers and implemented when the occasion arose are shown in Table 5.2. Both teachers were initially anxious that in developing their work in this manner they would be over-structuring it and perhaps might inhibit spontaneous interest. Both had their fears allayed by the enthusiasm of the children's response to new kinds of science experience that had not been a feature of their earlier approach.

Science as separate lessons

Teachers using this pattern will be devoting a regular amount of time to science and in one sense there is no problem of fitting science into the curriculum. But, and it can be a substantial but, if we apply the Chapter 3 checklists (page 15) to the situation, then a number of 'no' answers will present themselves because of the difficulty of developing investigative problem-solving science in short fixed periods of time. Children need ample time to discuss ideas, time to put these ideas into practical action whilst they are fresh in their minds and time also to make 'mistakes' and modify their work

Anticipated happening	Potential for sustained enquiry	Some action questions
Conkers craze (Junior classroom)	Investigating the hardness of conkers Making something to measure the hardness of conkers	Which method is best for hardening conkers? Can we make a hardness tester?
Christmas preparations (Infant classroom)	Cooking special things	What will happen if we use more/less of an ingredient in a mixture? Which (kitchen implement) is best for stirring?
	Packing presents	Which wrapper is best for sending a parcel through the post? Which fabric is best for Santa's sack?
	Decorations	Can we make something with – flashing coloured lights? – interesting shadow shapes?
	Toys	Can we invent a toy that will – make a noise? – move and change direction?

Table 5.2   Developing 'as and when' science

accordingly. This cannot be achieved in short lessons. Whilst not implying that science lessons are completely inappropriate, it is suggested that some rejigging of the class timetable may be needed if present practice is a weekly hour or so of science.

Teachers facing the problem have produced differing solutions. One infants teacher, having gained some experience of investigative work, decided that a science day would be more appropriate than her timetabled weekly one-and-a-half hours. Two years on she still operates a science day and supports it, against challenges that it means less time for basic skill work, on the grounds that the day involves much maths activity and an enormous amount of language development.

Others, particularly junior teachers, have found that merging two subjects can achieve greater time flexibility and is a powerful strategy for developing the work. Table 5.3 records some examples of subject mergers that have been used by teachers, developing from a separate subject approach. Initially each merger was in-troduced purely as an organisational device to provide flexitime for experimenting with new ways of working. Subsequently, the experience gained led to mergers with other subject areas and the increased curriculum integration provided even greater time flexibility.

Science as part of a wider topic

Organisationally, this curriculum pattern is a splendid one for developing science because:

- it usually involves children in finding out things for themselves and differing work may be going on in the classroom at any one time;
- in the planning teachers will normally seek links with all subjects including science.

Nevertheless, in many classes the so-called science component of a topic may not be science at all. What is called science is often no more than children using books to get information about scientific things. That is valuable activity for developing higher reading skills but it is *not*

Merger	Integrating topic	Examples of science component	Other subject focus
Science and maths	Ourselves	Who sees best? Is it true that big hands are stronger than small hands?	Graph work
Science and geography	Street study	Which shop sign is easiest to see? Shall we make a working model of a traffic light?	Mapping skills
Science and history	The Victorians	What was washday like without modern materials? — do detergents make washing easier? — which is the best place to dry wet clothes?	Empathy with people of the past
Science and art and craft	Camouflage	How will the colour change if we . . .? Is it true that background colour and pattern affects the way we see things?	Colour statements
Science and PE	Playground games	Which ball is the best bouncer? On which surface is it easiest to skip?	Small games skills

Table 5.3   Some examples of subject integration

scientific activity – the children are not handling and investigating materials. Even when a topic does include practical work, children's activities are not necessarily investigative; some might be better called 'cookery', when children follow instructional recipes to obtain an expected result.

So when we consider how to develop science in a topic approach we must consider how to ensure that appropriate skills-based work is included. This can usefully be achieved by re-orientating thinking about the planning of topic work and one example of such reorientation is shown in Fig. 5.1.

Resourcing and organising the work

These are major areas of demand in science and it is as well to acknowledge the fact. There are no universal procedures for instant success, but there is plenty of evidence that the tasks gets easier with experience.

Three common problems arise when we think about developing science beyond an early 'have-a-go' stage. These 'take-heart' comments about each one are intended to ease concern and practical suggestions for action follow later:

Familiar Planning

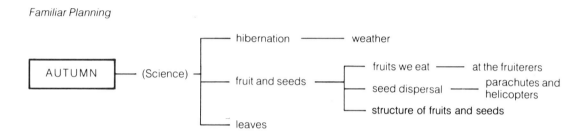

Reoriented Planning
SCIENCE COMPONENT (AUTUMN TOPIC)
Fruits and Seeds

Observing	Predicting	Investigating	Design and make
Sorting by properties, e.g. spinners, hookers, etc. Identifying flower- fruit-dispersed seed sequences. Do large flowers produce large fruits? Do all fruits have the same number of different parts? Do all fruits of the same kind have the same number of seeds? Can different kinds of . . . be distinguished by touch and smell?	Will its movement be the same if – . . .? – . . .? How many seeds will germinate? What will happen to the fruit by – next week? – . . .? Will it germinate earlier/later if – . . .? – . . .?	Does the movement alter . . . old/young? dry/wet? small/large? damaged/ undamaged? etc. Is it true that – birds like red berries? – all soft fruits stain? Which stains are easier to remove? Which fabric collects most hooked fruits? Which method is best for preserving . . .?	A machine to sort and grade . . . (windfall apples?) A machine to extract juice (and collect it) (and measure it) (and give an alarm if the collecting container overflows). A machine to pick . . . from out of reach places.

Fig. 5.1   Planning the science component of topic work

'There's hardly any science equipment in our school and I'm not sure what to get with the little money available.'

You need little, if any, specialised science equipment to develop from the have-a-go stage. In the longer term more will be needed.

'The work is so different from what the children have done before. I worry about noise, mess and possible lack of control.'

A lively and at times somewhat chaotic response is initially quite common. But as children settle to new ways of working, noise levels fall and things become more orderly.

'Science takes so much time and energy. I can't keep going at this rate or other work will suffer.'

Early work will inevitably be demanding because, like the children, teachers have to settle to new requirements. Pressures ease as everyone becomes more familiar with what is needed. The two basic resourcing questions are:

What do I need?
How do I store it?

At a more detailed level:

How do I select and get what I need without excessive cost and effort?
How do I organise the materials for effective learning?

We will consider resourcing in terms of realistic provision and effective storage so as to offer guidance to teachers who have 'had a go' and now wish to tool-up for developing science in a year's programme. We will not consider in detail all resources that ideally could support science. Teachers wanting such information will find help in the Science Resource Book produced by the Learning through Science Project (see Appendix 4).

Realistic Provision

It's useful to think of the materials children will need in three broad categories:

everyday materials likely to be used whatever the topic for investigation;
special equipment likely to be required for most topics;
topic-related materials.

Everyday materials

These will be needed for design-and-make activities and for inventing fair-testing procedures.

A useful starter list is:

boxes, tins, bottles, tubs in a variety of materials and sizes;
string, cotton, thread;
paperclips, drawing pins, sellotape;
plasticine, straws, marbles, elastic bands;
suitable adhesives.

Fortunately such materials are easy to obtain and they will be accumulated as activities develop.

Special equipment

This will be needed to extend the children's own capabilities, helping them to observe more closely and enabling them to measure things.

As a core collection obtain:

some means of magnifying;
equipment for measuring
time,
length, mass, volume,
temperature.

The exact nature of the equipment will depend on what's available in school. Most likely there will be maths equipment for basic measurement; magnifiers and thermometers are less common. If you lack these and cannot afford both, magnifiers will be the most useful investment.

Topic-related materials

These will be needed to stimulate children's interests and ideas and for particular investigations developed in the topic. This area of resourcing is likely to be the one that will need developing most strongly and we offer a well-tried common-sense strategy for approaching the task.

1. Decide on the topics your class will investigate during the coming year.

2. For each topic:
   (a) list relevant things that children could examine to excite their interest. This will become the topic collection;

(b) use a skills-scan to get action questions for children's activities. Anticipate and list the materials they will need to investigate them.

3. Assemble the materials! The exclamation mark is included because these three words mask the considerable effort needed and it is at this point intentions can get lost in practical reality. For this reason it is helpful to enlist parent help. Many teachers have been amazed at the response to letters of the kind: 'We will soon be studying ... We need ... Can you help?' Experience shows that requests for specific items gain most response, though it is always worth adding 'and anything else you think might be useful'.

Planning along these lines will lead to a considerable increase in resources and so attention has to turn from 'How do I get them?' to 'How do I store them?'

Effective storage

How and where science materials are stored will clearly depend on things such as money, space and the physical layout of a classroom. In deciding what will work best in the circumstances, it is useful to keep certain points in mind.

It helps if the everyday materials needed for testing and design-and-make activities are visually accessible to children.

Science activities have a strong inventive element, frequently involving the need to improvise. Being able to scan visually for possibilities helps children put their ideas into practical action, since they are not as adept as adults at 'seeing in the mind's eye'.

Way of achieving this include:

using silhouettes of the contents on the outside of storage boxes;
storing, as appropriate, in open-mesh containers;
having a low-level 'you may need' table for small items;
avoiding closed-cupboard storage.

It helps to use functional labels on storage containers. For example, 'length measurers' is more effective than 'rulers and tapes'; 'see-through things' communicates better than 'transparent objects'. This may seem a small point but it is quite important. Its effectiveness probably relates to the fact that investigative work requires children to become familiar with what materials do and what they themselves can do to the materials. Cataloguing resources by descriptive name does not assist this process.

It helps if topic-related materials are temporarily 'boxed' for easy distribution and retrieval.

In addition to collection boxes likely to be housed in a store cupboard, it is useful to consider boxing the materials that groups will need for investigative work over a period of a few weeks. The contents of these boxes will be determined by anticipating what children will need to respond to planned action questions. This arrangement will ease organisational problems, but it is not recommended as a long-term strategy for reasons identified below.

It helps if children take responsibility for aspects of resourcing.

Responsibility for everyday materials can be given to 'storekeepers', whose job it is to keep things tidy and report when materials need replenishing. One infants teacher had fun labels for the children to identify particular responsibilities (for boxes, tubes, etc.).

More importantly, work will develop more purposefully when children take increased responsibility for the choice of materials they use. We will return to this point on page 34, but in the present context it means:

making sure children are aware of the resources available to them, including those in the teacher's cupboard and in other parts of the school. Lists and simple catalogues will be needed and children can help in their compilation;
having 'house rules' that enable children to move freely and sensibly to get the materials they need.

These storage considerations are useful for shaping decisions, but there is one final point worth noting. As experience grows storage arrangements are likely to change and so, even if money is available, early investment in a storage system is not recommended. Most systems lack flexibility for coping with the variously sized objects you will accumulate and classify for use. Cardboard boxes are ideal for early-stage resourcing and they remain so for many teachers.

Avoiding chaos

Science work is inevitably noisy work because we want children:

- to talk freely among themselves since this is necessary to shape their ideas;
- to work investigatively with materials.

In many investigations the materials themselves will increase the noise level, as when weights fall in a strength test. Additionally, some happenings will so engage the children that they will react excitedly. We cannot escape such working noise. There will also be an element of mess because:

- everyday materials needed for investigation may well look untidy if stored accessibly and when in use as intended they may become an aesthetic eye sore;
- trial and error investigations will produce discarded items that temporarily add to the visual clutter and require attention in the clearing-up phase;
- liquids, however carefully handled, will sometimes get spilt.

These requirements understandably deter many teachers from developing genuinely investigative science. There is concern about accepting and coping with noise and mess and anxiety that things may get out of hand. Those worries are not groundless but, as is evident in classrooms where teachers are experienced in investigative work, commitment to the approach is not the same as commitment to chaos. How then to reduce its possibility?

It is necessary to accept that we cannot have a silent and immaculate classroom if we want to develop science work. Nevertheless, we do want noise and mess kept within reasonable bounds and there are strategies that are useful; some

self-evident, some more subtle.

If the work is new to the children it helps if:

- early sessions do not involve too many dramatic happenings;
- resources can be obtained by the children with minimum fuss (see previous section);
- we increase gradually the number of children involved, at any one time, in testing and design-and-make activities.

Less obviously — and crucially — it is important that we have clear ideas about how to ease children into new ways of working.

When the children are not used to working with materials in an investigative manner and are then expected to do so they may be uncertain about the changed demands being made on them. Uncertainty causes confusion and confusion can lead to unwanted off-task activity. This negative chain of events can be avoided if we help children make the transition, particularly in relation to their understanding of practical tasks and their use of materials.

In many situations the changed demands will be substantial, as indicated below.

Usual experience	New expectations
Told what to do	Expected to work things out for themselves
Given materials to work with	Expected to select appropriate materials themselves

The change is too great for most children to cope with unaided. Nor can many cope with it quickly. We can ease the transition by not expecting children to work unaided on very open problems until they have built up the appropriate experience.

Many problems, for example the question 'What . . . is best for . . .?' will initially be beyond most children's capabilities (unless they develop ideas through discussion with their teacher). Unaided groups will need to develop experience along the lines shown in Table 5.4.

The progression from 1 → 3 may take some children a term or more, but the sequence will help them to handle changed expectations and so will reduce the possibility of disorder.

Children's transition difficulties	Sequence of teacher help		
	1 ⟶	2 ⟶	3
Understanding what needs to be done	Do this (full ⟶ instructions given)	Try doing (guidance ⟶ with 'clues')	Invent a way to find out
Knowing which materials to use	Use these ⟶ (all necessary materials identified)	Use some or ⟶ all of these (some choice introduced)	Use what you need

Table 5.4

Keeping going

Initially the development of science will be demanding, but the rewards are great. The following advice comes from teachers who have been through this stage and now organise investigative work as a matter of course:

- don't expect too much too soon;
- don't try to do too much too soon;
- have a phased plan for developing the work;
- give a high priority to children taking decisions about their work;
- make use of supportive resources such as TV programmes and work cards, but don't become dependent upon them;
- share your experience with colleagues.

That is sound advice and the thrust of this chapter has been directed towards helping you take organisational decisions and practical action in developing the work.

There are, however, three notable omissions that need comment.

- We have not provided detailed suggestions for organising the children, though many of the suggestions will help you take appropriate decisions. We believe that this crucial aspect of organisation must be determined by individuals working within their own levels of confidence.
- We have not given detailed suggestions about the selection and organisation of content. We believe that at an early stage questions of content are not as important as a concern for ways of getting a programme of skill-related work established. In the longer term this matter has to be considered and we turn to it briefly in Chapter 8.
- We have not provided guidance about safety. This is an important concern underlying all aspects of organisation and is covered in Appendix 3.

Finally, as preparation for development, you might like to consider the activities you identified in the 'getting ideas' task of Chapter 4. How, in the light of this present chapter, could you resource and organise them for classroom action?

6. Where are we going?

- *Why is it worthwhile making the effort to start science and/or to keep it going in the ways suggested in the past few chapters?*
- *Isn't it enough to have, for instance, a good display and some posters and let the children get the rest from books?*
- *Why all the stress on providing materials for children to do things with?*
- *Why not let them follow instructions, so that they get things rights?*
- *Where is this way of working leading us?*

Some of these questions may well have occurred to you. You might have been asked them by other teachers or by parents. You need some good answers, ones which convince both you and others and provide a sound basis for present and future decisions about science work in your class and school.

This chapter is about *why* we are proposing working in the way we do in this book. We believe the purposes are sound, but they are not necessarily simple. If you feel your priority is to get on with the 'doing', you may not wish to do more than skim this chapter on first reading and spend more time trying the activities suggested in earlier chapters. But do come back to it later, because understanding the learning you are trying to nurture in children is a necessary basis for development of the practical aspects of the work. You'll find you can more easily create and use ideas of your own, once you know where you are going!

How are classroom decisions made?

Think for a moment about how you plan activities in some other area of the curriculum, say, PE. What makes you decide on certain activities rather than others, how to organise the children (in teams, groups or individually), whether or not to use certain equipment, whether to en-

courage or discourage competition? Your answers probably make reference to shortage of resources or space, but beyond that, where there is a real choice, they no doubt reflect what you want children to learn and how you think that learning is best brought about, for your children, at this time.

Making decisions in science is no different; we base them on what we think is worthwhile learning in science. Consider some of the ways in which science could be provided:

- teacher demonstration to the whole class;
- TV programmes;
- working from books only;
- working with materials individually;
- working with materials in groups;
- or in any other way you can think of.

Ask yourself to what extent does each of these give opportunity for children to:

(a) gather evidence at first hand by direct observation;

(b) use their own ideas in making sense of new experiences;

(c) explore and test materials;

(d) learn about others' ideas;

(e) make and test predictions based on their own and others' ideas;

(f) improve their ways of exploring, investigating and making sense of things around them.

It soon becomes apparent that certain kinds of provision preclude certain kinds of learning experience. If we value these then we have to organise provision that is consistent with them. Similar points could be made about the role of the teacher. For instance, which of (a) to (f) above are encouraged if the teacher takes the role of:

- provider of expert knowledge?
- fellow enquirer?
- technician and source of material?

This important relationship between classroom

decisions and learning opportunities has two implications:

- what we decide to provide and how we organise it constrains the kind of learning that can take place;

and, conversely,

- how we view learning will influence what opportunities we provide.

Let's look at these implications a little more closely.

Although we perhaps don't consciously carry out planning as a sequence of steps, starting with nothing except ideas, it is useful to think of the classroom situation as being a result of deciding what learning opportunities we want to provide. These in turn are a result of having in mind a certain view of what kind of learning is desirable.

In other words we *start* with:

1. an idea about the *kind of learning* we want to bring about (e.g. rote learning or learning with understanding);

2. then we decide the *learning experiences* which we think will be likely to give opportunities for this kind of learning;

3. next we arrange for these learning experiences to take place; this involves thinking about:
   materials and the roles of teacher and pupils in interacting with them
   the necessary organisation

4. finally, we decide how to *evaluate* how successful we have been.

Of course all of these decisions are made within a variety of *constraints* (limitations of money, time, energy and our own professional expertise). We never reach the position of being entirely satisfied with what we are doing and we should always keep it under review. Chapter 3 has given some ideas for reviewing the provision of activities aimed at improving the kind of learning we value. Chapter 7 suggests how we may find out whether the learning is taking place.

The four points listed above describe decision-making in a way which applies to any teaching that is internally consistent and well planned, whatever the type of learning which is embraced. It applies equally to rote learning as it does to more progressive views of learning. For example, take the teacher who was asked why she arranged her class with desks all facing the front, in rows, forbade the children to talk to each other and occupied them mainly in copying from the blackboard. She replied: 'Because that's the way they learn, from me and the blackboard, not by talking to each other.' However much we may disagree with her view, we must admit that her organisation and the experiences she provided were consistent with it!

However, our purpose here is not to justify rote learning! The example has been given only to emphasise that classroom decisions determine the kind of learning that children can experience. Therefore to show where we are going through the sorts of classroom activities we have been discussing and trying, it is necessary to think about children's learning.

Learning as changing ideas

Observations of children tackling new problems show the importance of making links between what is already known and what is new. Children faced with a new object or material invariably make statements or ask questions which show the struggle to make sense of the new in terms of existing ideas: 'I think it's . . .' or 'Look, it's just like . . .' are typical first reactions.

Similarly, adults search around to relate a new object or event to something encountered previously. For example:

Imagine being handed a type of fungus that you have never seen before. By looking at certain features you would soon recognise it as a natural rather than a human-made object and you might even be able to link experience of other fungi to suggest at least the sort of organism that it is.

When such previous learning is absent, we experience the feeling of not being able to make sense of something new — for instance, our first encounter with a computer! Then we, like children, need to build up understanding bit by bit, starting with something which *does* make sense to us and then linking in each new step to existing knowledge.

Children, like us, don't develop ideas from scratch about a new event or object, but learning begins by linking existing ideas to it and using them in an attempt to understand it. The ideas which are linked are checked to see if they are useful, if they fit further experience.

The linking process involves the skills of:

observing
hypothesising (attempting to explain)

and the checking processes involve the skills of:

raising questions
predicting
investigating
communicating

These skills are called science *process skills*. To complete the list that we are concerned with we should add the technological skills of:

designing and making

As we mentioned in Chapter 5, it is the practical testing and investigation that distinguishes science from other areas of the curriculum. Other areas also use observation, hypothesising, communication etc., but practical checking of predictions by changing some things and not others is particular to science. These checking processes are of special importance in children's experience of scientific activity.

The process of checking may show that the idea which has been linked *is* helpful, or it may show that it is not, or that it could be helpful if changed. So an idea may emerge from the checking as follows:

- reinforced after being shown to be useful in explaining something new;
- changed into something more useful;
- rejected (in which case another idea will have to be tried and checked).

Which of these things happens, i.e. what learning takes place, depends on the ideas the children have access to and the way in which the 'processing' is done.

Children's existing ideas

These are a mixture of partially formed scientific ideas, probably already changed by experience,

and ideas we might call 'everyday' rather than scientific, since they have very limited value in helping understanding. There are plenty of 'everyday' ideas around for children to pick up, from the media (especially advertisements — 'Nothing tastes better than butter!') and other parts of daily life. Often everyday language does not help, for children may not realise that we are speaking metaphorically when we talk about 'a living flame', 'feeding plants', etc.

As we don't want children to be forever limited to the ideas they already have it is necessary to broaden the range of ideas that they have access to. It is important to do this, however, in a way which acknowledges that the children will have their own ideas and which does not devalue them. We pick up this point again later in this chapter.

Children's process skills

These, too, can be limited and unsystematic. Take for instance some children who noticed that blocks of smooth, varnished wood stuck together when wet, and immediately described the wood as 'magnetic'. They were ignoring some available evidence in doing this (e.g. there were ways in which the blocks did not behave as magnets — there was no repulsion, only the sticking together, and for this the blocks had to be wet). They might have dismissed the idea if they had systematically considered the common features of magnets they had seen before and the wooden blocks. It was the *process* of linking the scientific idea of magnetism (very useful in understanding other experiences, but not this one) to the experience, not the idea itself, which was at fault in this case.

So, it is not difficult to imagine that an appropriate idea can be rejected when it should not be or an inappropriate idea could be accepted, depending on how the checking is done.

For example, in the fabrics activity in Chapter 2, children will probably have used ideas from their different past experiences to suggest which is the best fabric for 'keeping us dry':

'This one, because it's thickest' (the thicker the better).

'This one, because it's like my raincoat' (and mine keeps me dry).

'This one, because it's that sort of stuff that most raincoats are made out of' (so it must be best).

Now suppose the children tested the fabrics by comparing the thickness only. Then they would gain little information which would help them in answering the questions posed, for they would have made a comparison which was not relevant to it.

More generally, this is a common error of processing, where there is failure to identify the relevant variable to measure or compare. An everyday example is deciding which article to buy on the basis of the 'free offer' that comes with it, rather than on which is really more suitable for our purposes.

Another type of error in processing would occur if the children *were* testing the fabrics with water but did this by pouring different amounts onto the fabrics so that fair comparison was not possible. The answer would also be unhelpful in answering the problem.

Whilst in these cases it may not matter if the children decide that fabric *B* rather than fabric *C* is the 'best', it isn't difficult to think of instances where the consequences can have a longer-term effect on children's ideas. Imagine children setting up two simple circuits:

  one with red wires and a low-powered battery;
  one with blue wires and a higher-powered battery.

By comparing the brightness of the bulbs, it would be possible to conclude that the blue wires made the bulb brighter, unless the children realised that they were not making a fair comparison.

Developing ideas and using process skills

Our reasons for emphasising the use of science process skills is not, then, just for their own sake. Yes, it is good for children to be able to test things and check ideas in a fair way, but just as important a reason is that this is the only way in which they will build up useful ideas or concepts.

We will say a little more about what these useful concepts are in Chapter 8. For the moment let's keep the focus on the process of change in children's ideas, a process in which children use and check their ideas. The effect of the level of the children's development of skills on the development of their ideas can be seen through a few examples. For the sake of simplicity, we will consider the restricted list of skills suggested in Chapter 4.

Observation

Low-level development of this skill is characterised by global rather than detailed observation, attention to what is expected rather than what is actually there and a greater attention to differences than to similarities. Children whose observation is characterised in this way may easily miss the detail which enables them, for instance, to distinguish old bricks from new ones, to notice patterns in the way bricks are laid, to detect the signs of weathering on walls. They may not notice that certain shapes recur in buildings, that very large and very small buildings have some structures the same and others different — all observations which in themselves provide evidence allowing ideas to be checked.

Investigating (fair testing)

When objects or materials are compared one with another then not only are the materials being tested but there is an idea being checked — that adding more sand to the concrete mix will change the strength of the concrete when set, or that this kind of wood is more bendy than that kind.

Young children, or those at an immature level in fair testing skills, may not keep the amount of water or of other ingredients in their two concrete mixtures the same; they may compare a thick piece of one kind of wood with a thin piece of the other; they may, indeed, make a comparison which does not relate to strength at all — using appearance or hardness as a basis for judgement. What happens to their ideas about the composition of concrete or the properties of wood is not difficult to imagine.

Design-and-make skills

These skills enable children to try out ideas

about how things work or are constructed to achieve certain purposes. If these skills are as yet at an early stage of development children may, for instance, build a model which is more a symbol of a house rather than a representation of one in miniature. They might start from a cardboard box and paint on the bricks and roof, cutting out the windows and doors, etc. It doesn't matter in such a model that the roof is 'made' before the walls or how the bricks are arranged.

Contrast the way in which their ideas about foundations, the structure of walls, strength of structures, etc., would be developed in this symbolic building with what would happen if their model was a more representational one, as suggested in Chapter 4 (page 24).

As in all skills, the early stages in designing and making are a natural part of development and quite appropriate for younger children and those just beginning such activities. What is important, however, is that we recognise this as just the start and take deliberate steps to see that there is progression in children's skills.

Prediction

Using this skill gives children the chance to see if the patterns they have noticed, or the ideas they have already used in one situation, will apply and help in another one. It has a central part in helping the children to verify or modify their ideas.

When the skill is not well developed children's 'predictions' are close to 'guesses,' not really arising from ideas based on evidence and therefore not taking a large part in the development of those ideas. A child whose prediction skills are more advanced will have a reason for suggesting that, for instance, the concrete will set more quickly if warm water is used and testing his prediction will be helping him check his idea. These are important reasons for helping children from the 'guessing' stage to more advanced stages in the skill of prediction.

What we want is for children to obtain evidence which helps them to change their initial, limited ideas into more widely applicable and helpful ones (see Chapter 8). We can see that this change depends on *how* they obtain the evidence, as well as on *what* their ideas are and the nature of the problem.

The role of process skills in the development of children's ideas in science is the reason why it is necessary to emphasise skills in science activities and to foster attitudes which control their use. The attitudes include *respect for evidence* (willingness to collect and use evidence), *flexibility* (willingness to change ideas in the face of evidence) and *critical reflection* (the habit of reflecting on and critically reviewing ways in which an investigation has been carried out).

Learning with understanding

The argument that learning involves *changing* ideas also helps to explain why it isn't enough to tell children how things work or provide information from books. This is because:

facts which do not link into existing ideas don't affect the way children really understand the world around, any more than learning rhymes and jingles by heart;

children still use their *own* ideas in explaining things to themselves, even though they may be able to recite the 'right' answers they have been given;

to bring change in the children's own ideas, we have to help them to change these ideas for themselves by realising that they do not fit evidence, or not as well as alternative ideas do. We cannot do the changing for them. All we can do is to help them in the process by providing opportunities for ideas to be tried out, challenged, stretched, changed or replaced.

We mustn't forget, of course, that we are concerned with young children, whose experience and ways of thinking are limited. This means that:

if we rely too much on their narrow range of existing ideas they may not be able to understand new experience;

it is important for them to realise that there are other ideas besides their own to be tried.

This is where discussion is so important, for it exposes children to what others think and to a range of suggestions to try out beyond the ones they themselves have thought of.

When children begin to realise the advantages of considering alternative ideas, they should be able to look for these in books and other sources of information. These sources should not be offered as giving the 'right' answers, rather as providing suggestions that are worthwhile considering. So we need to provide the materials, the talk, the right kinds of questions and, above all, the supportive atmosphere in which they can expose and change their ideas without being made to feel they were 'wrong'.

Bringing all this together, we can use the decision-making framework described on page 38 to summarise the experiences, organisation, roles and evaluation criteria for this kind of learning.

1. *View of learning*

Learning is through children:

- making their own sense of experience;
- linking to existing ideas and past experience;
- changing ideas to fit evidence better.

2. *Learning experiences*

Experiences in which children:

  actively seek evidence through the senses;
  check ideas against evidence;
  take account of others' ideas;
  seek more effective ways of testing ideas.

3. *Materials and ways of interacting with them*

The role of materials being:

- to provide evidence and means of testing predictions;
- to arouse curiosity and stimulate exploration and investigation.

The role of the teacher being:

- to find out existing ideas and help children test predictions based on them;
- to help children devise and reflect upon ways of testing predictions fairly;
- to promote interaction with materials and others' ideas.

The role of the children being:

- to become involved in the discussion of ideas and in making predictions and proposing ways of testing them.

Organisation

Such as to:

- optimise opportunity to interact with real materials;
- optimise access to a range of ideas from other children, adults, books, media.

4. *Evaluation criteria*

The criterion for learning opportunities being:

- the extent to which the children have the experiences indicated in the questions on page 15 of Chapter 3.

The criterion for children's learning being:

- their progress in developing skills, attitudes and concepts as discussed in Chapter 7.

Where we are going?

Hopefully we can now see the purposes of the kinds of experiences we are suggesting. The summary on page 44 brings together the reasons in terms of children's learning for the suggestions made about organising classroom work.

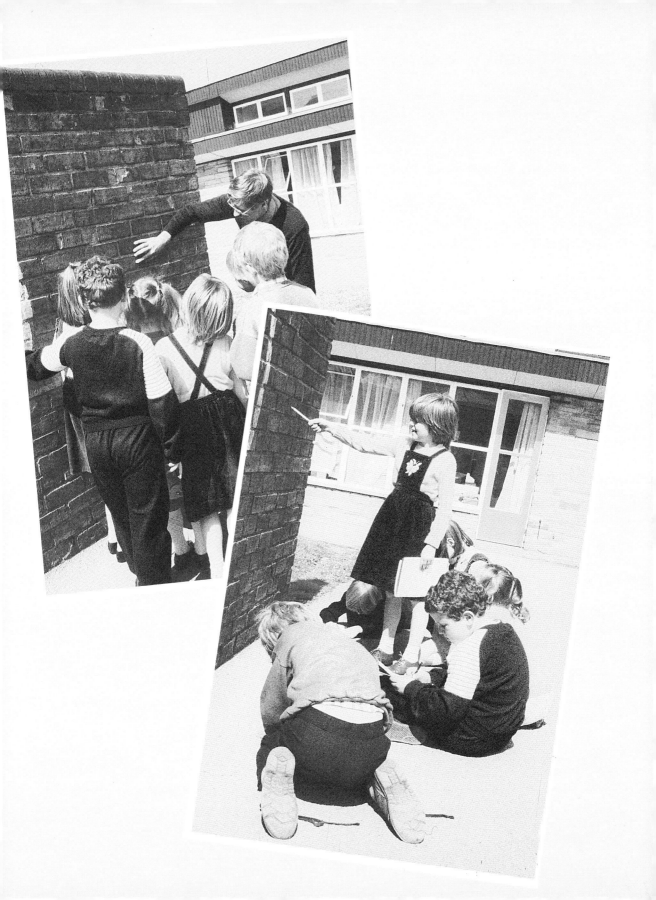

What to do	*Where it leads*
Provide materials and arrange for children to interact with them and with other real things in the environment.	Children gather evidence using their own senses and through their own activity, check their ideas against this evidence of how real things behave.
Organise activities to encourage discussion in small groups.	Children hear other ideas than their own, refine their own through explaining them and have a range of suggestions to try.
Discuss with children in small groups or individually.	Children talk about their ideas so that these are evident to themselves and the teacher; are encouraged to check their findings, to think critically about what they have done and how they have done it; teachers offer ideas and direct children to sources of information if relevant.
Hold whole-class discussion and reporting sessions.	Children record/report for a purpose; listen to and discuss what others have done; realise that there are different approaches (and results) from theirs.
Make available books, displays and pictures in the classroom and give access to sources of information outside.	Children compare their ideas with others'; gain information to extend or change their ideas, possibly to raise questions leading to further enquiries.
Teach the conventions used in graphs, tables, charts, etc. and the techniques of using measuring instruments.	Children are able to use quantitative techniques when they wish to; can increase the accuracy of measurement; can choose appropriate forms of communication.

7. Development in learning

Making a start is fine, but to build on it productively we have to consider the progress we want children to make and how we can help it to take place. When discussing children's learning in Chapter 6 we talked of the role of *process skills* and related *attitudes* in the change and development of *concepts*, making the points that:

- process skills are mental and physical skills used in 'processing' information about objects, events and materials that children encounter;
- unless these skills develop from immature, 'everyday' forms to more mature and scientific forms then children are unlikely to develop scientific concepts with understanding.

So one of the main tasks in teaching primary science is to help children develop the process skills; indeed all the classroom ideas we have suggested in this book have been chosen with this task in mind. Even with these suggestions, however, you will be in a much better position to help children in this development if you know what course it is likely to take. In this chapter, therefore, we are concerned with describing the changes in skills and attitudes, and with trying to do this in a way that will be of use to you in finding the point of development of your children. The related question of the development of concepts is taken up in the next chapter.

As you probably realise, the subject of this chapter represents a shift in focus from teacher activities to what the children are doing. If you feel comfortable with this and ready to make this change in your concerns, it is a sure sign that your confidence is growing!

As the word 'development' comes up so often in this discussion it might be useful to say something about its meaning to prevent possible misunderstanding.

The meaning(s) of development

In one sense development implies an automatic process of change which will take place as a matter of course — the kind of 'development' which happens when a caterpillar changes into a butterfly. In another sense it implies a more active process, as when we talk of the 'development' of a programme of activities — by no means a spontaneous set of events! In both senses a series of changes is implied, but the supposed mechanism of the changes is different — internal in the case of the butterfly and external in the other case (programmes of activities do not develop themselves, unfortunately).

We are using the word here to mean the series of changes which take place in children's ideas, skills or attitudes. These changes are neither wholly spontaneous, taking place as a result only of maturation, nor wholly externally determined, taking place as a result of 'training'. The exact way in which the changes do take place is not known, but experience suggests that it is through a mixture of maturity and experience. There is also evidence that, in science, children's ideas and skills could develop (i.e. pass through various changes) more quickly if children were given more opportunity for appropriate experience. But this isn't to say there is no limit set by mental maturity to the development.

In plain terms, what this means is:

- don't sit back and wait for ideas and skills to develop with time. That won't happen.

On the other hand:

- don't expect to be able to accelerate children's progress through the developmental changes to an unlimited extent; change takes time.

One of the reasons for this, as suggested in Chapter 6, is that these changes have to be brought about by the children themselves.

Although no-one can change the ideas or skills of another, it is the job of teachers to provide the conditions and encouragement for children themselves to make the changes which constitute development.

Describing development

In the following pages there are sets of questions about things children may or may not be able to do relating to science skills and attitudes. The questions for each one are arranged roughly in order of development. The order cannot be exact. Uncertainty about the sequence in development is inevitable given individual differences in experience and response to experience; but they give a general guide.

You might use these sets of questions in three main ways:

(a) to help you to identify the level of development of the children in your class;
(b) to help in pitching activities at the right sort of level to provide encouragement for skill development;
(c) to help in observing the progress of individual children and in catering for the variation between children in the point of development they have reached.

Finding the general level of your class

1. Look through each list of questions, bearing in mind the general way in which the children responded to a particular activity — perhaps the fabrics testing.

2. In each list find the point where your answers stop being a firm 'yes' and gradually change into 'no'.

3. You will probably find this more difficult to do for some lists than for others. Don't worry about this. Perhaps deal first with the 'short list' of skills as introduced in Chapter 4 (observing, investigating, designing and making, and predicting), then come back later to the others.

Observing

This is the skill of taking in information about all the things around. It can, and should at various times, involve the use of all the senses. Development in observation has two aspects which at first seem conflicting; on the one hand the increasing ability to notice as much detail as possible and not be limited by what one expects to find, and on the other hand the increasing ability to distinguish between what is relevant to a particular problem and what is not. There is a danger in narrowing the focus of observation too soon to what is regarded as relevant. This is a long-term aim and priority at the primary level should be given to attention to detail and sequence.

Do the children:

- succeed in identifying obvious differences and similarities between objects and materials?
- make use of several senses in exploring objects or materials?
- identify differences of detail between objects or materials?
- identify points of similarity between objects where differences are more obvious than similarities?
- use their senses appropriately and extend the range of sight using a hand lens or microscope as necessary?
- notice patterns, relationships of sequences that are to be found in a series of observations?
- distinguish from many observations those which are relevant to the problem in hand?

Explaining (hypothesising)

The skill of explaining something involves using previous experience and existing ideas. The ideas which are applied are selected because of some similarity between the new event or objects and those encountered previously.

Some of children's explanations are no more than statements of coincident circumstances; others take the forms of labels (as if giving something a name explains it — typically, a word such as 'energy' is used in this way). More advanced explanations are in terms of mechanisms, but of course there is always a further question to be asked about any explanation. (See Chapter 8 for some discussion of the levels of explanation appropriate at the primary level).

Do the children:

- attempt to give an explanation which is consistent with evidence, even if only in terms of the presence of certain features or circumstances?
- attempt to explain things in terms of a relevant idea from previous experience even if they go no further than naming it?
- suggest not only what but how something is brought about, even if the 'how' would be difficult to check?
- show awareness that there may be more than one explanation which fits the evidence?
- give explanations which suggest how an observed effect or situation is brought about and which could be checked?
- show awareness that all explanations are tentative and never proved beyond doubt?

Predicting

Making a prediction goes beyond the evidence which has been gathered and uses what is available to suggest what will happen after some process has continued or changes have been made. The use of evidence or previous experience distinguishes a prediction from a guess, which does not appear to have a rational basis.

There are various degrees of using evidence, which signify levels of sophistication in using this skill. At the lower levels, children tend to jump to conclusions which have only a slight link with the evidence. At a rather more advanced level the link is firmer but perhaps still intuitive. Later still comes the ability to explain how the evidence is used in arriving at the prediction by some form of extrapolation or interpolation.

Do the children:

- attempt to make a prediction relating to a problem even if it is not derived from the evidence?
- make some use of evidence in making a prediction, rather than basing it on preconceived ideas?
- make reasonable predictions which fit the evidence without necessarily being able to make the justification explicit?

- explain how evidence has been used in making predictions?
- perceive and use patterns in information or observations to make justified interpolations or extrapolations?
- show caution in making assumptions about the generalisation of patterns beyond the evidence available?

Raising questions

Whilst we are concerned here with questions of a particular kind we do not want to imply that these are the only worthwhile questions. Children should be encouraged to ask all kinds of questions, since it is through doing this that they can help form links between previous and new experience and so enlarge their understanding. In science, however, we are concerned with questions which can be investigated, that children can answer through action — action which involves the science process skills.

Fundamental to the development of the skill of raising investigable questions is the gradual recognition of the kind of question with which science is concerned. This is basic to the appreciation of scientific activity. Beyond the primary years the pupils may learn to recognise some questions as philosophical, others as essentially matters of value or of aesthetic judgement. In the primary years, however, the distinction is between those which cannot be answered through scientific enquiry (e.g. why does someone prefer one pattern of wallpaper to another?) and those which can if they are expressed in certain ways (e.g. which fabric is best for making a raincoat? is testable when we have decided what 'best' means and how to test it).

Do the children:

- readily ask a variety of questions which include investigable and non-investigable ones?
- recognise a difference between an investigable question and one which cannot be answered by investigation?
- realise when an investigable question is in a testable form?

generally, in science, ask questions which are potentially investigable?

- quite often express their own questions in testable form?

- ask questions which arise from making a prediction or giving an explanation that can be tested?

Investigating

What is involved here is both planning and carrying out a series of actions which are related to finding an answer to a particular question. It will generally involve action to change something and then observing the effect of the change whilst other things are kept the same; or it may involve comparing different things when treated in the same way. Key features are: selecting the appropriate variable to change or objects to compare, keeping all other things the same (controlling variables), and observing or measuring the relevant effect systematically and carefully.

Children at the early levels of this skill have to 'think as they go' in their investigations; those at more advanced levels will be able to plan ahead before starting an investigation, to anticipate what is involved and take action to avoid possible problems.

Do the children:

- start with a useful general approach even if details are lacking or need further thought?

- have some idea of the variable that has to be changed or what different things are to be compared?

- keep the same the things which should not change for a fair test?

- have some idea beforehand of what to look for to obtain a result?

- choose a realistic way of measuring or comparing things to obtain the results?

- take steps to ensure that the results obtained are as accurate as they can reasonably be?

Communicating

This is the skill of giving others, as clearly as possible, an indication of ideas and evidence used in and arising from an investigation. It also involves receiving, understanding and responding to, information given by others.

Informal communication, both written and spoken, is a means of sorting out ideas and linking different experiences to each other. Formal communication skills are quite heavily dependent on the knowledge of conventions, such as names of objects or events, how to draw graphs, tables, charts and use symbols. Development in communication skills doesn't just reside in knowing the conventions, but rather in the appropriate use of the various means of communication to suit the receivers and the type of information.

Do the children:

- talk freely about their activities and the ideas they have, with or without making a written record?

- listen to others' ideas and look at their results?

- report events coherently in drawings, writing, models, paintings?

- use tables, graphs and charts to record and report results when these are suggested?

- regularly and spontaneously use information books to check or supplement their investigations?

- choose a form for recording or presenting results which is both considered and justified?

Designing and making

Design-and-make skills, at their simplest, are used and developed through constructing 'things that work'. Such things 'work' in that they provide a solution to a real problem; they do not necessarily move or have moving parts, though many will have. A model bridge built high enough to allow boats to pass under and strong enough to allow loads to pass over, 'works' in this sense as much as things which may be made to move across and beneath it.

There is an interconnected range of skills which depend partly on knowledge (of use of tools and of properties of materials, for example), partly on development of broader concepts (such as structure/function relationships) and partly on creative thinking (to think up possible solutions to try).

Do the children:

- make models to represent real things even if the resemblance is only symbolic rather than structural?
- choose appropriate materials for constructing working models?
- use tools effectively and safely?
- succeed in making simple models that work?
- consider the structure/function of materials used in real-life examples when designing a model?
- realise the discrepancies between a real object and a working model which are necessitated by limitations of materials and scale of construction?

Attitudes

Attitudes are distinct from skills and concepts in that they describe the willingness to act or react in a certain way, whilst skills and concepts describe the required know-how to do so. However, it is clear that in practice these three are much less distinct, for on the one hand, we can't really be predisposed to use a skill or idea that we don't possess; on the other hand, there must be some willingness to use the skill or concept in order to develop it at all.

It is useful, though, to regard attitudes as separable from skills and concepts to some degree. It helps us pay particular attention to them and reminds us that not only do we need to teach children so that they *can* use scientific skills and ideas but also so that they *will* use them to work scientifically.

Attitudes are 'caught' as much as 'taught'; they are encouraged most effectively in children by a mixture of example and approval of the behaviour they describe. So, to foster respect for evidence, flexibility of ideas and critical thinking in our children, the best way is to show these attitudes in our own behaviour.

Willingness to collect and use evidence (respect for evidence)

Do the children:

- report results which are supported by evidence even if the interpretation is influenced by preconceived ideas?
- realise when the evidence doesn't fit a conclusion based on expectations, although they may challenge the evidence rather than the conclusion?
- check parts of the evidence which don't fit an overall pattern or conclusions?
- accept only interpretations or conclusions for which there is supporting evidence?
- show a desire to collect further evidence to check conclusions before accepting them?
- recognise that no conclusion is so firm that it can't be challenged by further evidence?

Willingness to change ideas in the light of evidence (flexibility)

Do the children:

- readily change what they say they think, though this may be due to a desire to please rather than the force of argument or evidence?
- change ideas when there is considerable evidence against the existing ones and little in their favour?
- show willingness to consider alternative ideas which may fit the evidence, even if they prefer their own in the end?
- relinquish or change ideas after considering evidence?
- spontaneously seek other ideas which may fit the evidence rather than accepting the first which seems to fit?
- recognise that ideas can be changed by thinking and reflecting about different ways of making sense of the same evidence?

Willingness to review procedures critically (critical reflection)

Do the children:

review what they have done after an investigation even though they may only justify rather than criticise it?

consider some alternative procedures which could have been used without necessarily realising their advantages and disadvantages?

discuss ways in which what they have done could have been improved even if only in detail?

consider, when encouraged, the pros and cons of alternative ways of approaching a problem to the one they have used?

initiate review of a completed investigation to identify how procedures could have been improved?

spontaneously review and improve procedures at the planning stage and in the course of an investigation as well as after completion?

Encouraging skill and attitude development

1. Make sure there is plenty of opportunity for children to do the things they *can* do (the yes answers) in deciding further activities and in carrying out your role.

2. Give the children opportunity and encouragement to try those things next down on the lists, where they do not yet succeed. Try this within activities where the subject matter is familiar and non-threatening (i.e. *not* in an activity which involves using equipment never used before, but perhaps in extending investigations using already familiar materials).

3. Listen and watch how the children tackle the parts of activities which require advancement of a skill or attitude. Discuss with them the problems they have and find out their view of the task. This will almost certainly give you some clues as to how to help them make progress in these areas.

Observing and helping the progress of individuals

When you have become more familiar with the questions in the lists and with the development they describe, you will find that you begin to carry this knowledge around in your head. You will then be able to use it in noticing how individual children go about their work.

1. Apply one set of questions to an individual child. It is best to do this using observations in more than one activity and noting any variations in the 'yes' and 'no' answers which apply.

2. Focus attention on what is described in questions for which yes/no answers were given or on the next in the list after consistent 'yeses'. Make sure there is opportunity for the child to try these in future activities.

3. Discuss with the child the response to aspects of the activity which demand advancement of the skill or attitude and listen to his or her view of what is done. You will probably then be able to decide whether the child is on the point of the change which is required or whether more practice and consolidation at an earlier point of development is appropriate.

4. Repeat for the other sets of questions.

In practice it is unlikely that you will proceed one set at a time in a step-by-step fashion. Any one activity gives information about several, sometimes all, of the skills and attitudes and you will find that you are picking up information which helps in finding the level in several skills at the same time.

It is worthwhile, though, to check systematically that you have the necessary evidence for making a decision about each list. If you find that you haven't the evidence, then it may be that the opportunities of the child in relation to a skill or attitude are rather limited. Then you might revisit Chapter 3 and review the children's activities.

8. Development in teaching science

Working with wider perspectives

In the previous two chapters we have discussed children's development and introduced frameworks for thinking about how it can be recognised and encouraged. Now, in this final chapter, we focus on the further development in teaching that will take the work beyond the stage of establishing a foundation skills-based programme.

If you have been actively responding to the preceding chapters you may have experienced some changes in your approach to science. Some of the changes recorded by other teachers after similar experience were expressed as follows:

'I'm more confident.'
'I'm more willing to give children their head and work with their ideas.'
'It's made me reflect more on what I am doing.'
'I'm more conscious of the importance of good questioning and my questioning style has changed.'
'One of the important changes for me is to alter the way I organise materials.'

These are heartening comments and may reflect aspects of your own development. Collectively, they paint a picture of a teacher who has come to grips with early problems of making a start in science and is now developing it well.

There is no clear-cut distinction between the have-a-go stage and the later development stage, for the 'first-stage' concerns are continuous ones that become easier to handle with experience. At the same time, as provision of skills-based work progresses, it is likely that questions will emerge that were not evident earlier. Two common examples are:

should I just concentrate on skills in science, even though I plan work in other areas (e.g. maths) with concepts in mind?

does it matter that the children have investigated certain topics but have no experience of others?

Attacking such concerns is what we mean by working with wider perspectives, and for practical reasons we can identify two areas for development:

deciding how to extend the planning of science to give attention to concepts and content as well as to children's process skills;
thinking about how the work we do with one class fits into the wider context of a whole school approach.

Concepts and content

Is there a difference between concepts and content?
The short answer to this question is 'Yes, there is'. There is a useful distinction and one which becomes crucial in progressive stages of curriculum planning. In brief:

content refers to the subject matter of an activity or topic; the particular object, materials and events that it concerns (building a model house with wood and cardboard; seeing which fabric is most waterproof; observing the feeding, growth, movement and reproduction of snails, etc.);

concepts refer to the ideas which these activities may help children to build; ideas which have a wider relevance than just to the particular content studied (e.g. about the properties of materials; the needs, behaviour and life cycle of living things).

In terms of these examples it is evident that the concepts relating to materials could have been developed by doing other things than building a model house or testing fabrics with water and that living things other than snails could have been studied to help develop the overall ideas about living things. This point can be generalised to say that a concept or idea can be developed through many different activities or topics, that is, there are several content routes to the same concept. The importance of this relationship is that teachers can choose the routes and content to take advantage of their particular children's environment and interests and still cover the same concepts in their science work.

Three further points about the relationship between content and concepts emphasise the importance of not confusing the two.

(a) Whilst we can with certainty say that children have encountered certain content (have made a pulley to lift a load, for example) we cannot be equally sure that an idea about force and movement has been grasped.

(b) A concept, being a generalisation (see the next section), has to be encountered by children in a range of content. One particular content is unlikely to be sufficient to establish a concept: there must be range and variety in the content made available.

(c) The content of any one activity or topic is very likely to provide opportunity for several different concepts to be developed.

These last two points together mean that there isn't a simple one-to-one relationship between content and concepts, but something much more complex. This becomes clearer when we pursue the nature of concepts a little further.

About concepts

Some questions which almost always arise in discussing science concepts at the primary level are:

- what are concepts in any case, and are science concepts any different in kind from others?

- how do science concepts develop and how does teaching help?

- which concepts — and to what level — should we be concerned with in primary science?

What are concepts?

Although more complex definitions can be given it is helpful to think of concepts in terms of what they *are* and what they *do*.

They *are*

generalisations of some kind concerning the similar features of different objects or events.

For instance, if we have a concept of 'chair', then on entering a room for the first time we recognise certain objects in it as chairs even though we may never have seen chairs of that shape and design before. The 'chair concept' comprises the essential qualities which are the same for all chairs and enables us to focus on certain 'chair' properties and ignore other features in classifying the objects.

What concepts *do*

is to enable us to use past experience in dealing with new experience.

We couldn't cope with everyday life without this ability — every object and event would present an overwhelming overload to our senses.

Concepts which relate together a wide range of different phenomena are generally described as 'high-level' and tend to be more abstract. Those which have a restricted range are 'low-level' and are more concrete.

Here are some examples of concepts that we use daily. Try putting them in order from high to low.

money
coin
inflation

Science concepts — a sub-set of all concepts — are

those generalisations which help us to understand the order in the natural and physical world around.

They concern the patterns that can be discerned in the way things behave and in materials. Like all other concepts they can differ in range and degree of abstraction.

Also, like other concepts, they sometimes take the form of a single word, such as:

animal	sound	dissolving
plant	reflection	floating

and sometimes the form of a relationship for which there is no single word, for example:

'whenever there is a sound there is something vibrating';

'a complete circuit of conducting material is needed for an electric current to flow'.

We have to be careful when dealing with single-word concepts because:

1. the label, the word, cannot be found from investigation; it has to be taught. But knowing it doesn't necessarily indicate that for a child the word brings with it the full meaning that it has for an adult. The word 'animal' is an example; at first it is used by children to mean only furry, four-legged mammals, not including fish, worms, spiders, etc. Later the word will take on a wider meaning;

2. there are various levels at which a concept can be understood and the use of a single word cannot distinguish one from another and so can easily be misunderstood.

How do concepts develop?

A good deal has been said about this in Chapter 6. Without going over the same ground again, it is important to emphasise our view of learning and the relationship between processes and concepts.

Children will already have some ideas or concepts which they bring to a new experience in trying to understand it. Thus learning involves *change in concepts*, not creating them from scratch.

Concepts change and develop by being checked against evidence from new experiences. The checking involves the processes of observing, explaining, predicting, investigating, etc.

How the processing is done has a profound influence on the emerging ideas; process skills have an important role in concept development.

These points reiterate something about the mechanism of change, but they do not tell us what sorts of change are taking place. Is there a pattern in children's changing concepts? How do the concepts of a typical six-year-old differ from those of a typical ten-year-old?

There certainly are patterns in the development of concepts and it is useful to know them, as long as we resist the temptation to say that because a child is x years old then she or he 'ought' to have these kinds of ideas rather than those.

The two most helpful patterns to have in mind concern *range* and *degree of abstraction* of concepts.

Taking *range* first, what we find is that:

the concepts of young children are limited to a few objects or events; they might be called mini-concepts ('what I eat for breakfast', 'what I eat for dinner');

gradually the mini-concepts become linked up into concepts encompassing a wider range of instances ('things I eat');

the process of merging continues so that broader concepts are formed ('food').

We might represent the process as shown in Fig. 8.1 opposite.

If only mini-concepts are available, there is less chance of new experience fitting in than if the larger concepts exist. For instance, the concept of food is more likely to help explain what is happening when a thrush is seen consuming a snail, than a concept of 'what I eat for breakfast'. So broader concepts have greater power in helping to explain new experience. Consequently, teaching should aim to:

help children to make links between ideas they already have, but may keep distinct;

broaden the range of their ideas by applying them to new experience and making the necessary changes to accommodate it.

The pattern in *degree of abstraction* indicates that:

young children's concepts or mini-concepts are related to what is immediately observable to the senses ('strong things, like bridges are made of strong materials like stone, brick and cement');

gradually they are able to generalise about relationships between things ('strength depends on the way a structure like a bridge is built as well as the materials used');

concepts become related to more abstract qualities ('structure/function relationship').

54

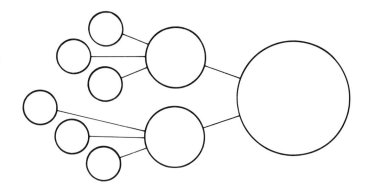

Mini-concepts limited in range and thus in usefulness for explaining new events

Wide-ranging concepts which have greater use in explaining a range of new events

Fig. 8.1

This sequence indicates a change in the nature of the entities which are related together in the concept. To be able to deal with abstract qualities requires a great deal of experience of dealing with concrete things and forming generalisations about them, and the general mental ability for abstract thought. Children in the primary years, by and large, do not reach the point of being able to use and develop concepts at a high level of abstraction. They can, however, be helped to consider the way things behave, not just what they are, and to begin to categorise things in terms of relationships.

What concepts and at what level?

Given a commitment to children learning with understanding through using process skills, then we must be concerned with concepts which:

1. relate to children's immediate experience;
2. they can use, check and change through active investigation.

One list of concepts which fits these requirements is reproduced in Appendix 1. It suggests some one word concepts (italicised) which are appropriate, and fits these into relationships in an attempt to make clear the level which is intended. The concepts are those which children might have developed by the end of the primary years, given a thorough experience of investigative science throughout this period.

To develop a firm grasp of the items in the list a great deal of exploration across a wide range of experience is required. For example, both the fabric activities of Chapter 2 and the school building activities of Chapter 4 provide experience relating to the concepts under the heading 'Materials and their properties and uses'. It isn't essential for these particular activities to be used – many others would help the development of the same concepts. But any one set of activities is not enough on its own; others are required to consolidate the concepts under the one heading.

When you add together the activities which might contribute to all the concept areas, sufficiently to give children chance to develop ideas through their own activity, the result is more than enough to fill the curriculum time for science throughout the primary school.

Expressing concepts as relationships, as in Appendix 1, helps to answer the question:

how far do we go with each concept?

What the list is saying is, for instance in relation to sound:

'If the children, through their exploration of sound, understand that there is something vibrating whenever there is a sound, then that's enough.'

We don't need to teach that the higher the frequency of the vibration the higher the pitch of the sound, or what wavelength is, etc. They may explore ways of making fast and slower vibrations and notice the change in the pitch – that's fine – but it should be seen as consolidating the main idea that the sound is caused by vibration. This is the idea to keep firmly in view. A whole range of experience is needed before children really grasp this as something which always happens, or at least as something to which they can't find an exception.

Similarly, many teachers have reported that children don't immediately distinguish eggs from excrement, or relate a flower to the seeds which it leads to and from which it grows. Such evidence shows how much experience and bringing together of different ideas is needed for the appreciation of life processes and the place of an egg or seed in the life cycle of some living things. So the simple ideas suggested under the heading 'Ourselves and other animals', which are:

> 'There is a wide variety of different living things called animals; different kinds feed, grow, move, protect themselves and reproduce in different ways ... Animals of the same kind go through the same life cycles'

require a great deal of experience and discussion of that experience. Again, we should see the aim as being the development by children themselves of these simple ideas and not more complex ones about the digestive system, cell structure or the like.

What the list may help you to do is to see that, at the same time as helping children to develop science skills and attitudes, their activities are also helping them to build basic concepts which they can use in understanding further experiences. Where this fits into the overall scientific development of your children will of course depend on their age and experience and on what is happening in other classes in the school. For the moment we will go no further, for these touch on matters which take us beyond the concerns of one class to the role of science in the school as a whole.

Pointers for action

It will be evident from the previous section that we should not ignore concepts and content. However, there is some poorly-charted territory to cross before we turn conviction into practical action. Lists of appropriate concepts and content help us to know where we want to go but getting there is another matter. The concept/content issue is a difficult area for all teachers, however experienced, because a common and detailed concern for it is comparatively new on the primary science scene. Don't be deterred by this, rather regard it as a 'take-heart' comment.

What, then, needs doing? As a first step it helps to think about provision for developing concepts and decisions about content selection as separate issues even though the two are closely intertwined and will be so in practice.

Helping children in developing concepts

Bearing in mind the points made earlier about concepts (page 54), we can help children develop their ideas by:

> *Providing opportunities for them to tell us what they really think about things*

Much of our verbal interaction with children in skills-based work will be directed towards helping them sort out working procedures. To tap their ideas we need to focus more on the reasons they have for the suggestions they make and the things they choose to do and on their comments on what happens during investigations. Judicious use of reason-seeking questions will help.

'Why do you say that?'
'What makes you choose ...?'

> *Analysing what children are likely to mean by the things they tell us, especially when their comments appear strange*

Consider, for example, events in one classroom, where the children were engaged in a topic on eggs. They carried out some excellent skills-focused investigations which included:

> looking closely at the content of a hen's egg;
> seeing what changes take place when an egg is cooked;
> finding out which box was best for packaging eggs.

Their teacher had egg shells of other birds which the children examined, making comparisons with a hen's egg. When asked if these eggs would be the same inside as the hen's egg, there was an emphatic 'no' from one lad. Probing, the teacher asked 'Why do you say that?' Answer: 'Because this one (the gull's egg) will have fish-bones.' Unexpected comment and scientific nonsense, but clearly a statement with real meaning for the child. We might interpret the comment as

> 'Gulls eat fish (I've seen them swooping into the sea and that's what I think they are

doing) – the fish goes inside the gull and gets into the eggs (that's what I think happens) – fish have bones – so there will be fish bones in the gulls' eggs (that's my understanding).'

Such an interpretation tells us very clearly that the boy has, as yet, little scientific understanding of life processes. His comments indicate a lack of understanding of the 'egginess' of eggs and their significance in the life cycle of animals. We cannot develop that understanding in the child overnight but we can log the fact of his ideas as something that other children are likely to share and use the knowledge to shape future work. Certainly we could note that more work on life cycles would be useful.

  Organising class discussion in which we show that we value children's contributions, however odd, and through which children can be helped to analyse their own interpretations of things and consider alternatives

Discussion is extremely valuable in science work because it helps children:

  clarify their thoughts in order to communicate ideas to others;
  recognise that others may have different interpretations of shared experiences;
  appreciate the need for evidence to support a point of view.

Such experience can emerge in small group discussions but it is enriched by the range of contributions that is made possible by whole class discussion.

But discussion does not just happen. If we accept its value we need to take action on two fronts.

1. We must monitor our responses to contributions. How often, feeling pressed for time and over-keen to get on with what we've planned, do we respond with a semidismissive 'y-e-s' and quickly draw in someone else? Children are not slow to pick up the implied irrelevance of their contribution and they may be less forthcoming in future.
2. We must make sure that children have the chance to acquire the necessary confidence and skills to engage in discussion. This may require conscious development as part of an oral language programme.

  Increasing our ability to plan work 'through the mind of a child'

In planning it is important to anticipate both what the children are likely to do and what they are likely to think about the things they investigate.

As an example, take planning work on soil. We need first to identify what children are likely to know about soil and the likely meanings it has for them. Here are some infants' responses to the question 'What can you tell me about soil?'

  'It's dirt.'
  'Seeds grow in soil.'
  'Worms live in soil.'
  'My dad digs it.'
  'Our dog digs it too.'

Interestingly, some of these comments were made also in a lower-junior and an upper-junior class.

With such anticipated responses as a starting point we can consider how ideas might develop to build up a wider concept of soil. Figure 8.2 shows how one teacher attempted to map ideas in this way, starting with the anticipated response 'Worms live in soil.' Other starting ideas would produce different detail but would have a similar pattern. The advantage of planning is that it enables us to consider links between statements and ask ourselves what could children do to help make the links themselves? (For example, a useful experience would be for the children to investigate 'Which soil is easiest to tunnel through?') In this way planning for development of ideas and skills goes hand in hand.

You may find the approach daunting, possibly because you feel you don't know enough about soil. But look carefully at the statements in Fig. 8.2 overleaf. They are largely general-knowledge statements, emerging from the analysis of soil and its properties and interaction with things in it.

Why not 'have a go' yourself, changing the topic from soil to water?

The same strategy applies for any topic and can be summarised as:

1. Identify children's likely knowledge and meanings.

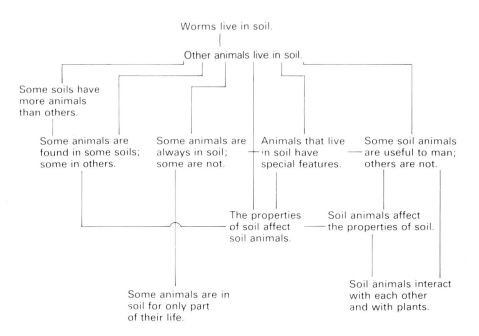

Fig. 8.2   Concept map used in planned part of a soil topic 'through the mind of a child' (Juniors)

2. Map a possible development of ideas.
3. Consider what experience children need to link one idea to another.

We may find it difficult to extract the 'meanings' from children, but the attempt will heighten our awareness of the concept aspect of the work.

Decisions about content

Picking up a question raised at the beginning of this chapter,

> 'Does it matter that the children have investigated certain topics but have no experience of others?'

the most appropriate answer is:

> 'no, when making a start, but *yes* later'.

This needs some explanation. In terms of purposes, science work has two *interacting* strands: skills and concepts. Neither develops in isolation from the other and though we may discuss them separately and put different emphasis on one or other at times, their interaction in learning must not be forgotten. With this firmly in mind, it is helpful to concentrate initially on skills and later to give concepts more attention whilst maintaining attention to skill development. This has some implications for content selection, remembering the relationship between content and concepts mentioned earlier (page 52).

There always has to be some content and to say that, in the initial period, skills are emphasised does not mean we can assume content does not matter. What it does mean is that the content is chosen to give opportunities primarily for process-skill development and later it is chosen to provide opportunities for concept development as well. In the initial stage, then, the content of what the children investigate should:

1. interest and motivate children and teacher;
2. have rich potential for problem-solving activities that promote a wide range of skills;
3. be such that it can be adequately and safely resourced.

These are early-stage criteria for selecting content and as such supply the *no* part of the answer to the question at the beginning of this section. All these criteria remain appropriate as science work develops but then, with an increased concern for concepts, we need to add to the list. As work progresses, in addition to the above, the content should:

4. provide children with experience in all the broad areas of science;
5. enable children to develop ideas within each area and encourage their appreciation of the relationships between them.

Acceptance of these criteria turns our answer into *yes*.

Let's consider points 4 and 5 more carefully, starting with broad areas. The argument is basically simple. Just as we want children to have a nutritionally balanced diet to grow well, so we want them to encounter a range of content that will contribute to a balanced diet of important ideas in science. We don't want to stunt their conceptual growth by, say, working only with living things and neglecting the physical side of science. So it is important to identify the areas of content that will shape planning in science. Earlier we referred to a list of concepts, which is reproduced in Appendix 1. These are what we wish children to develop through the primary years and to do so they must encounter content in various broad areas. The National Curriculum for science for England and Wales proposes 12 content themes for the primary years which are also listed in Appendix 1.

In considering this list, it is important not to let the label conjure up a recollection of old-style syllabus headings with its connotation of facts to be learnt. Better to think of them as categories of content to be handled in an investigative manner.

You may worry that defining broad areas of content in this way may stifle spontaneity and so conflict with the idea of selecting content on the basis of interest and motivation. This possibility is unlikely to appear in practice if we accept a role of 'interest-generator' to complement an earlier emphasis on content in which the children have already shown an interest. In doing so we may need to monitor the influence of our own interests and motivation.

Many teachers find the 'Forces' and 'Energy transfers' categories worrying. The words appear more 'scientific' and you can't give children a chunk of force or energy to handle physically; there is an anxiety that more scientific knowledge will be needed than when dealing with living things and materials. Consequently, motivation can be low and the area receive scant treatment. If this is your concern, try thinking of force as the 'push and pull' of things and energy as the 'go' of things. There are lots of push/pull, go/no-go things around to help you get going.

The last of the criteria for content selection, point 5 above, also needs some comment. When we say that, over time, children should be able to develop ideas within each broad area we are not suggesting that they should encounter all possible content. That's clearly impossible. But it is important to view each area in terms of its potential breadth and depth. By breadth we mean the range of content it contains; depth relates to the development of children's ideas.

In practical terms, a concern for breadth means considering the wide range of things that could be investigated and not restricting the work to a few 'favourites'. For instance, here is one teacher's programme for living things:

Autumn	fruits and seeds
	hibernating animals
Spring	buds, seeds, and bulbs
	tadpoles
Summer	trees
	fish

The work, however enjoyable, lacks breadth because the children's encounter with living things only involves vertebrate animals and flowering plants. There is a wealth of other life forms that could be investigated and they are neglected for no better reason than habit.

Concern for depth means consciously thinking about how we can help children develop their ideas along the line suggested on pages 56–8 and in doing so we will identify relationships between areas. For example, the action question 'Which soil is easiest to tunnel through?' (page 57) emerged from thinking about animals (living things) moving through (force, movement and energy) soil (materials). Many relationships will come to light in this way but we can help their emergence if, when planning activities, we keep in mind certain important higher-order concepts which are embedded in the list in Appendix 1. This is how some of the statements on the soil map came into being. It's a powerful strategy, but like many things in this book, it needs practice to operate well. Early attempts may be frustrating, so let's finish with one last 'take-heart' comment. Rome wasn't built in a day, neither is an ideal science programme for a class!

The material we have included on content decision impinges on decisions of other colleagues in school, since the children will come with content experience and leave to receive more. The question 'How does what I choose to do fit in with what others do?' raises the important issue of continuity and progression in scientific experience throughout a school and, although crucial to a child, it is beyond the scope of this book. Sufficient here to remind ourselves that we shouldn't ignore concepts and content — thoughtful action, however hesitant, is better than no action at all!

References

DES (1978) *Primary Education in England.* London, HMSO

DES (1985) *Science 5—16: A Statement of Policy.* London, HMSO

Elstgeest, J. (1985) 'Encounter, interaction and dialogue.' In *Primary Science: Taking the Plunge.* W. Harlen (Ed.). London, Heinemann Educational

Harlen, W. (1985) *Teaching and Learning Primary Science.* London, Harper and Row

Jelly, S. 'Helping children raise questions — and answering them.' In *Primary Science: Taking the Plunge.* W. Harlen (Ed.). London, Heinemann

Nuffield Junior Science Project (1967) *Teachers' Guide 1, Apparatus,* and *Animals and Plants.* London, Collins Educational

Science 5/13 Project (1972—5) *Teachers' Guides* (26 titles). London, Macdonald

Appendix 1

Learning in science at the primary level: process skills, attitudes and concepts

Process skills

(For definitions and descriptions of development please see pages 46-9.)

Observing
Explaining (hypothesising)
Predicting
Raising questions
Investigating (fair testing)
Communicating
Designing and making

Attitudes

(For definitions and descriptions of development please see pages 49-50.)

Willingness to collect and use evidence (respect for evidence)
Willingness to change ideas in the light of evidence (flexibility)
Willingness to review procedures critically (critical reflection)

Concepts

(These are taken from W. Harlen, *Teaching and Learning Primary Science*, pages 78-9.)

Sight and light

Seeing things involves *light* coming from the objects seen into our eyes. Light passes from one place to another in straight lines but can be made to change direction if things are put in the way.

Hot and cold and temperature changes

When *hot* things cool down or *cold* things warm up there is a change of temperature which can be felt and measured by a thermometer. Things hotter than their surroundings cool down as they lose heat and things colder than their surroundings warm up as they gain heat. These changes can be slowed down by various devices and materials. Gaining or losing heat can change things (melting, freezing, evaporating, condensing).

Hearing and producing sound

Hearing things involves *sound* coming from them reaching our ears. Sound is created by objects moving rapidly (*vibrating*).

Movement and forces

Moving from one place to another takes a certain time; the shorter the time for the same distance the *faster* the movement. *Speed* is a measure of how fast something moves, usually how far it goes in a certain time, a second, minute or hour. *Force* is what tries to start or stop something moving or change its motion. When there is any change in motion there must be a force acting.

Air and breathing

There is air in the 'empty' spaces around us. We feel it only when it moves, as a *wind*. Air is a substance, called a *gas*; like all substance it has mass. Water, a *liquid*, can go into the air in the form of a vapour (*evaporation*) and comes out under certain conditions (*condensation*). Other things can mix with air; some can be detected by *smell* when they reach our noses. Living things need and use air.

How things behave in water

Some, but not all, things *dissolve* in water. Some, but not all, things *float* in water. Whether a thing floats or not depends on how heavy it is for its size.

Ourselves and other animals

There is a wide variety of different living things called animals; different kinds feed, grow, move, protect themselves and reproduce in different ways. They usually do these things in ways which suit them for living in particular environments. Animals of the same kind go through the same *life cycles*.

Soil and growth of plants

There is a wide variety of different living things called *plants*, different kinds feed, grow and reproduce in different ways. Many are green and produce the food they need through a process which needs light. Soil is a mixture of different things some of which are needed by plants to grow.

Sky, seasons and weather

The sun, moon and stars move relative to the earth in regular repeated patterns. Changes in the apparent positions of the sun in the sky are connected with night and day, and seasonal changes in the weather. Water condensing from the air under certain conditions gives rain, cloud, frost and snow.

Materials and their properties and uses

Materials are grouped according to their properties, such as whether they are *hard, bendy, transparent, strong*; different types are used for different purposes on account of their properties. The *strength* of structures made from a particular material depends on their form.

Simple electric circuits

Some materials allow an electric current to pass through them (*conductors*); others effectively prevent a current from flowing (*insulators*). There is always a continuous path of a conducting material when electricity flows from one terminal of a battery to the other.

Content themes

Proposals for the National Curriculum for science in England and Wales identify 12 content themes for the primary level. These are listed below on the left with linking lines showing how the above concepts relate to them.

National Curriculum themes	Concepts
The variety of life	
Processes of life	Ourselves and other animals
Genetics and evolution	Soil and growth of plants
Human influences on the Earth	
Types and uses of materials	How things behave in water
	Materials and their properties and uses
Earth and atmosphere	Sky, seasons and weather
Forces	Movement and forces
Electricity and magnetism	Simple electric circuits
Energy transfers	Hot and cold and temperature changes
Sound and music	Hearing and producing sound
Using light	Sight and light
The Earth in space	Sky, seasons and weather

Appendix 2
Handling children's questions

(From S. Jelly, 'Helping children raise questions − and answering them', in *Primary Science: Taking the Plunge*, pages 53−56.)

Spontaneous questions from children come in various forms and carry a variety of meanings. Consider for example the following questions. How would you respond to each?

1. What is a baby tiger called?
2. What makes it rain?
3. Why can you see yourself in a window?
4. Why is the hamster ill?
5. If I mix these (paints), what colour will I get?
6. If God made the world, who made God?
7. How long do cows live?
8. How does a computer work?
9. When will the tadpoles be frogs?
10. Are there people in outer space?

Clearly the nature of each question shapes our response to it. Even assuming we wanted to give children the correct answers, we could not do so in all cases. Question 6 has no answer, but we can of course respond to it. Question 10 is similar; it has no certain answer but we could provide a conjectural one based on some relevant evidence. All the other questions do have answers, but this does not mean that each answer is similar in kind, nor does it mean that all answers are known to the teacher, nor are all answers equally accessible to children.

When we analyse what we do everyday as part of our stock in-trade, namely respond to children's questions, we encounter a highly complex situation. Not only do questions vary in kind, requiring answers that differ in kind, but children also have different reasons for asking a question. The question may mean 'I want a direct answer', it might mean 'I've asked the question to show you I'm interested but I'm not after a literal answer.' Or, it could mean, 'I've

asked the question because I want your attention − the answer is not important.' Given all these variables how then should we handle the questions raised spontaneously in science work? The comment of one teacher is pertinent here:

'The children's questions worry me. I can deal with the child who just wants attention, but because I've no science background I take other questions at face value and get bothered when I don't know the answer. I don't mind saying I don't know, though I don't want to do it too often. I've tried the "let's find out together" approach, but it's not easy and can be very frustrating.'

Many teachers will identify with these remarks and what follows is a suggested strategy for those in a similar position. It's not the only strategy possible, nor is it completely fail safe, but it has helped a large number of teachers deal with difficult questions. By difficult questions I mean those that require complex information and/or explanation for a full answer. The approach does not apply to simple informational questions such as 1, 7 and 9 on the list above because these are easy to handle, either by telling or by reference to books, or expertise, in ways familiar to the children in other subject areas. Nor is it relevant to spontaneous questions of the productive kind discussed earlier, because these can be answered by doing. Essentially it is a strategy for handling complex questions and in particular those of the 'why' kind that are the most frequent of all spontaneous questions. They are difficult questions because they carry an apparent request for a full explanation which may not be known to the teacher and, in any case, is likely to be conceptionally beyond a child's understanding.

The strategy recommended is one that turns the question to practical action with a 'let's see what we can do to understand more' approach.

The teaching skill involved is the ability to 'turn' the question. Consider, for example, a situation in which children are exploring the properties of fabrics. They have dropped water on different types and become fascinated by the fact that water stays 'like a little ball' on felt. They tilt the felt, rolling the ball around, and someone asks 'Why is it like a ball?'. How might the question be turned by applying the 'doing more to understand' approach? We need to analyse the situation quickly and use what I call a 'variables scan'. The explanation must relate to something 'going on' between the water and the felt surface so causing the ball. That being so, ideas for children's activities will come if we consider ways in which the situation could be varied to better understand the making of the ball. We could explore surfaces keeping the drop the same, and explore drops keeping the surface the same. These thoughts can prompt others that bring ideas nearer to what children might do. For example:

1. Focusing on the surface, keeping the drop the same:
    What is special about the felt that helps make the ball?
    Which fabrics are good 'ball-makers'?
    Which are poor?
    What have the good ball-making fabrics in common?
    What surfaces are good ball-makers?
    What properties do these share with the good ball-making fabrics?
    Can we turn the felt into a poor ball-maker?

2. Focusing on the water drop, keeping the surface the same:
    Are all fluids good ball-makers?
    Can we turn the water into a poor ball-maker?

Notice how the 'variables scan' results in the development of productive questions that can be explored by the children. The original question has been turned to practical activity and children exploring along these lines will certainly enlarge their understanding of what is involved in the phenomenon. They will not arrive at a detailed explanation but may be led towards simple generalization of their experience, such as 'A ball will form when . . .' or 'It will not form when . . .'.

Some teachers see the strategy as one of diversion (which it is) and are uneasy that the original question remains unanswered, but does this matter? The question has promoted worthwhile scientific enquiry and we must remember that its meaning for the child may well have been 'I'm asking it to communicate my interest'. For such children interest has certainly been developed and children who may have initiated the question as a request for explanation in practice, are normally satisfied by the work their question generates.

The strategy can be summarized as follows:

Analyse the question
↓
Consider if it can be 'turned'
to practical activity (with its
'real' materials or by simulating them)
↓
Carry out a 'variables scan' and identify
productive questions
↓
Use questions to promote activity
↓
Consider simple generalization children
might make *from experience*

Appendix 3

Safety in primary science

(From C. Gilbert and P. Matthews, *A Guide to Primary Science Policy*, Edinburgh, Oliver and Boyd, page 19.)

Safety points

1. Liquids or objects (e.g. peas or marbles) spilt or dropped onto the floor may cause falls.
2. Objects placed in unaccustomed places during experiments may cause falls, or breakage of equipment.
3. Children should not run about when carrying equipment which might shatter (e.g. glass).
4. Glass should always be handled carefully and when possible plastic should be used in its place. Some objects, e.g. glass mirrors, may have their edges taped to avoid sharp cutting edges.
5. Care should be taken when holding objects close to the eye.
6. The ways in which germs can be transferred are many and the chances of this should be reduced by cleanliness of 'bodies' and equipment. Hands should be washed before touching things to be put into the mouth. Some things may need to be disinfected.
7. Children should not carry heavy loads, e.g. buckets full of water.
8. Some animals cause allergies.
9. Tasting of things should not be allowed except under strict supervision.
10. Children should know some plants are poisonous and they should be aware of what these are.
11. It may be felt the climbing of trees should be discouraged.
12. Extreme physical exertion may not be wise for some children.
13. Plastic bags cause suffocation and are potential hazards.
14. Use of cutting devices, knives, scissors, chisels, etc, can be dangerous. Children need to be shown the correct techniques for their use and should not be allowed to discover these.
15. Throwing projectiles or dropping things need to be done at carefully chosen and monitored places.
16. Even with household chemicals, e.g. Alka Seltzer tablets, vinegar, etc., care is needed and pressure can cause chemicals to shoot out. Safety goggles may be considered. Usually these are not essential but using them would be a way of encouraging good habits for later secondary chemistry.
17. With naked flames, e.g. lighted candles, children should be warned about long hair, ties and other bits of clothing not coming near the flame. Candles should be firmly fixed in stable holders.
18. Butane burners should not be used by children except under very careful supervision by a teacher who is always present during the experiment. The flames of some spirit burners are not easily seen.
19. Pupils should not look directly at the sun even through dark glass or plastic. Atmospheric conditions sometimes allow a clean view of the sun's disc with the naked eye. Even then, this should only be viewed for a very short time.
20. Pupils should not look at very bright lights, e.g. a projector beam.
21. Hot water should not be put into thick glass containers since they may crack due to uneven expansion setting up strains in the glass.
22. Thermometers should be used carefully, the bulb being very thin. If a thermometer breaks and mercury is spilt it should be

carefully gathered up (not by the fingers). In primary schools it should be disposed of by burying it in a place where the ground will not be disturbed.

23. Liquids which produce inflammable vapours (e.g. methylated spirits) should not be used by children.

24. Some glues can be hazardous both from the inflammability and inhaling point of view.

25. Pupils should use only low voltage supplies, e.g. torch batteries. Mains electricity should not be used for electricity and magnetism experiments in the primary school. Where a piece of apparatus powered from the mains, e.g. a computer or projector, is used then it should be connected and switched on under supervision. Do not leave apparatus connected and switched on unnecessarily.

26. Lenses can focus light and heat. Children should not look at intense sources of these through lenses or any devices with lenses, e.g. telescopes. Pupils must not look directly at the sun through lenses, especially through telescopes or binoculars.

27. Although not a hazard to pupils, magnets can affect other equipment, e.g. watches.

28. Even low voltages can cause short lengths of thin wire to become hot, even to glow and ignite things.

29. Animals obtained for keeping in school must be disease-free and are best purchased from accredited sources.

30. Such stock may, from dirty food and unsatisfactory cleaning, become carriers and transmitters of parasites.

31. The care of animals during holiday times should be considered very carefully indeed.

32. Animals must not be allowed to come into contact with wild animals and if a school has mice etc. special precautions should be taken.

33. Animals can bite and scratch and resulting wounds must be carefully treated.

34. People with cuts or infections should not come into contact with animals.

35. Wild animals, alive or dead, should not be brought into the school.

36. Food for animals should be carefully and safely stored so it does not become infested.

37. Moulds which have been grown should be carefully destroyed.

38. Convex lenses and concave mirrors can, in strong sunlight, cause fires if heat is focused onto something inflammable.

39. Objects should not be pushed into ears or noses.

40. Special care is needed when using matches.

Appendix 4

A selection of publications in primary science and technology

Books for teachers, other than classroom materials, published since 1980

A Guide to Primary Science Policy
Gilbert, C. and Matthews, P. (1984)
Edinburgh, Oliver and Boyd
> A short pamphlet by the authors of the *LOOK!* scheme which provides help for schools in constructing their school policy for science with reference to the use of *LOOK!* materials.

Switched Off: the Science Education of Girls
Harding, J. (1983)
York, Longman Group Resources Unit for the Schools Council
> Describes how girls can be deterred from science, examines the evidence for possible reasons and suggests a range of strategies for avoiding the problem, particularly at secondary level but also with relevance to primary practice.

Guides to Assessment in Education: Science
Harlen, W. (1983)
London, Macmillan Education
> Discusses in non-technical language the nature of primary science, the assessment of pupils and the evaluation of learning opportunity.

Primary Science: Taking the Plunge
Harlen, W. (Ed.) (1985)
London, Heinemann Educational Books
> Contains specially written chapters dealing with aspects of the teacher's role and giving practical advice for starting science, developing various science process skills and concepts and for dealing with children's questions.

Teaching and Learning Primary Science
Harlen, W. (1985)
London, Harper and Row
> Takes the nature of learning in science at the primary level as the basis for discussing content, methods, organisation, resources and the teacher's role.

Approaching Primary Science
Hodgson, B. and Scanlon, E. (Eds) (1985)

London, Harper and Row/The Open University
> Readings taken from previous publications and journals covering aims, issues, research findings and classroom practice. Some overlap with the selection in Richards and Holford's collection (see below).

We Make Kettles: Studying Industry in the Primary School
Jamieson, I. (Ed.) (1984)
York, Longman Group Resources Unit for the Schools Council
> A collection of case studies concerned with industry-related projects in primary schools, including some useful guidelines for making visits to industry successful with young children.

Formulating a School Policy: with an index to Science 5/13
Learning Through Science Project (1980)
London, Macdonald Educational
> Useful suggestions for how schools might go about developing a school policy for science. Includes a comprehensive index to the Science 5/13 Units.

Science Resources for Primary and Middle Schools
Learning through Science Project (1982)
London, Macdonald Educational
> Gives information and advice about the collection, construction and storage of equipment and other resources.

The Teaching of Primary Science: Policy and Practice
Richards, C. and Holford, D. (Eds) (1984)
Lewes, Sussex, Falmer Press.
> Short chapters by a number of authors, some articles having been published elsewhere. They provide background about the nature of primary science, theories of learning, approaches to curriculum development, issues of implementation and recent initiatives.

Girlfriendly Science: Avoiding Sex Bias in the Curriculum
Smail, B. (1984)
York, Longman Group Resources Unit for the Schools Council
> Creates awareness of the sex-bias that can creep into science teaching and gives practical

suggestions for action at the primary as well as secondary level.

Exploring Primary Science and Technology with Microcomputers
Stewart, J. (Ed.) (1984)
London, Council for Educational Technology for the Microelectronics Education Programme
A series of short chapers about pros and cons of using microcomputers and exploring their various roles in the teaching of science, electronics and control technology.

Unesco Handbook for Science Teachers
Paris, Unesco/London, Heinemann (1980)
Provides discussion, some theoretical background and suggestions for relevant science. Relates mainly to upper primary and lower secondary, where science is distinguished from other parts of the curriculum.

Design and Technology 5—12
Williams, P. and Jinks, D. (1985)
London, Falmer Press
Discusses, in plain terms and through examples, the nature of technology and its place in the primary curriculum. Gives basic practical help in terms of technology and case studies and guidance in curriculum planning in design technology.

ASE Publications

(All available from the Publications Department, Association for Science Education, College Lane, Hatfield, Herts AL10 9AA.)

Language in Science Study Series No. 16 (1980)
This book takes the form of a series of explorations into various aspects of language use. It discusses the crucial role that language takes in the learning process and how science can contribute to children's overall language development.

The Headteacher and Primary Science (revised 1981)
Discusses the role of the head in primary school policy and practice and suggests how heads can help to ensure appropriate opportunities for learning in science for all pupils in their schools.

A Post of Responsibility in Science (revised 1981)
A collection of ideas from teachers holding scale posts for science which may help others in similar posts to define and carry through their role.

Experiencing Energy (1980): Book 1 'Moving Things'; Book 2 'Burning, Warmth and Sunlight'; Book 3 'Working with Electricity'
Sources of ideas for classroom work in these topics.

ASE Primary Science (bound copies): Issues 1—9; Issues 10—18
Collections of classroom ideas supplied by teachers and well illustrated with children's work, originally published in leaflet form for subscribers.

Choosing Published Primary Science Materials for Use in the Classroom (1985)
Gives a description and evaluation of nine sets of classroom materials using explicit criteria drawn up by the Primary Science Sub-Committee of the ASE (now replaced).

Primary Science Review
A journal, published three times a year, containing articles, reviews, research findings and notices relating to primary science. Available to those subscribing as individuals or as primary school members of the ASE. For details, apply to the Membership Department of the ASE at the address given above.

DES publications (including APU), since 1980

A Framework for the School Curriculum
London, HMSO (1980)

Education 5 to 9: an Illustrative Survey of 80 First Schools in England
London, HMSO (1982)

9-13 Middle Schools. An Illustrative Survey
London, HMSO (1983)

Science in Primary Schools A discussion paper produced by the HMI Science Committee.
London, DES (1983)

Education 8 to 12 in Combined and Middle Schools. An HMI survey
London, HMSO (1985)

Science 5-16: A Statement of Policy
London, HMSO (1985)

Full APU Reports

Science in Schools. Age 11 APU Report No. 1
London, HMSO (1981)

Science in Schools. Age 11 APU Report No. 2
London, DES (1983) (out of print)

Science in Schools. Age 11 APU Report No. 3
London, DES (1984)

Science in Schools. Age 11 APU Report No. 4
London, DES (1985)

Science in Schools. Age 11 Review of the First Five APU
Surveys
London, DES (1988)

APU Reports for teachers

(All published by the DES and available from the
ASE Publications Department at the address given
on page 69.)

No. 1 *Science at Age 11*
       A summary of the findings of the first two
       surveys with implications for teaching.

No. 4 *Science Assessment Framework at Age 11*
       A illustrated account of the categories used
       for assessment in the APU surveys, with
       examples of questions and marking schemes.

No. 6 *Practical Testing*
       An account of the practical testing carried out
       at ages 11, 13 and 15 in the APU surveys.

No. 8 *Planning Scientific Investigations at Age 11*
       Describes how the skills of planning
       investigations were assessed in the APU
       surveys, presents the findings and discusses
       some implications for teaching.

Materials for use in the classroom

1. Teachers' guides

Science 5/13
London, Macdonald
       Gives activities, advice, examples and
       background information relating to a range of
       topics; children's activities being related to
       broad stages of educational development. Titles
       are:
       With Objectives in Mind: Guide to Science 5/13
       Early Experiences
       Change (Stages 1 & 2 and Background)
       Change (Stage 3)
       Children and Plastics (Stages 1 & 2)
       Coloured Things (Stages 1 & 2)
       Holes, Gaps and Cavities (Stages 1 & 2)
       Like and Unlike (Stages 1, 2 & 3)
       Metals (Stages 1 & 2)
       Metals (Background information)
       Minibeasts (Stages 1 & 2)
       Ourselves (Stages 1 & 2)
       Science from Toys (Stages 1 & 2)
       Science, Models and Toys (Stage 3)
       Structures and Forces (Stages 1 & 2)
       Scructures and Forces (Stage 3)

Time (Stages 1 & 2 and Background)
Trees (Stages 1 & 2)
Working with Wood (Stages 1 & 2)
Working with Wood (Background information)
Using the Environment:
    Early Explorations
    Investigations (Part 1)
    Investigations (Part 2)
    Tackling Problems (Part 1)
    Tackling Problems (Part 2)
    Ways and Means

Teaching Primary Science
London, Macdonald
       Nine units and an overall guide, particularly
       aimed at helping student teachers and those
       with little scientific training. Well illustrated
       activities relating to common topics:
       Candles
       Seeds and Seedlings
       Paints and Materials
       Science from Waterplay
       Fibres and Fabrics
       Mirrors and Magnifiers
       Science from Wood
       Musical Instruments
       Aerial Models
       Also Introduction and Guide to *Teaching Primary
       Science*.

Science for Children with Learning Difficulties
Learning Through Science Project
London, Macdonald
       Describes simple and child-centred activities
       suitable for infants children and older children
       with learning difficulties. Activities are related
       to topics taken from children's immediate
       surroundings and everyday experience.

Practical Primary Science
Showell, R. (1983)
London, Ward Lock Educational
       Consists of a series of activities for children,
       concisely described, with suggestions for further
       development and a resource list.

Teaching Science to Infants
Showell, R. (1979)
London, Ward Lock Educational
       Describes activities for infants relating to
       science-based topics.

A Source Book for Primary Science Education
Ward, A. (1983)
London, Hodder and Stoughton
       Contains a large number of starting points for
       activities for children preceded by a description
       of a view of the nature of primary school science.

2. Materials for pupils and teachers

LOOK! Primary Science Project

Gilbert, C. and Matthews, P. (1981–85)
Edinburgh, Oliver and Boyd
>Consists of three sets, each of 71 cards:
>>*A First LOOK!*(resource cards for teachers giving activities for 4 to 8 year olds)
>>*LOOK! Pupils' Pack A* (cards for 7 to 9 year olds)
>>*LOOK! Teachers' Guide A*
>>*LOOK! Pupils' Pack B* (cards for 9 to 11 year olds)
>>*LOOK! Teachers' Guide B*
>The cards present activities intended to provide a complete science course, but can be used selectively. They are printed on different coloured cards which relate to nine topics. They contain instructions for the children and extension work is suggested in the teacher's notes for each card. Children following these will have plenty of opportunity for activity but less for thinking and planning for themselves. The Guides give helpful advice about apparatus, safety and classroom use of the activities and some science backgound.

Learning through Science
London, Macdonald
>Twelve sets of workcards in full colour which can be used independently or in conjunction with *Science 5/13*. Each set contains two copies of 24 cards and a Teacher's Guide.
>Titles are:

Ourselves	Colour
Materials	Sky and Space
All Around	Out of Doors
On the Move	Moving Around
Earth	Electricity
Which and What?	Time, Growth and Change

>Also: Guide and Index (overall introduction and teacher's guide).
>The materials are suitable for children with a reading age of 7+ and provide open-ended activities, with a balance of suggestions and invitations for children to extend activities further. Teachers have to select activities suitable for children at different stages of development. There is potential for the development of a range of science process skills, attitudes and concepts.

Exploring
Brown, Christine, Brown, Christopher, Edwards, E., Roberts, A. and Young, B.L.
Cambridge, Cambridge Educational
>Consists of four units, each covering one year's work of the junior school and containing twelve modules. A module consists of a number of pupils' cards, a Teacher's Card and a card of extension activities. The content is related to eight themes and the cards can be purchased in theme sets as well as year units. The activities are designed to emphasise the scientific skills of observing, measuring, predicting, communicating, classifying, interpreting, hypothesising, experimenting, counting and problem-solving.

Science in a Topic
Kincaid, D. and Coles, P.
London, Hulton Educational
>Nine topic books and a Teacher's Guide. Designed to support and illustrate the teaching of science through topic work for 8 to 11 year olds. Material for pupils and teacher is combined in each topic book and is related to the rest of the curriculum. Titles are:

Ships	Houses and Homes
Communication	Clothes and Costumes
Food	Moving on Land
In the Air	Roads, Bridges and Tunnels
Sports and Games	

Science Through Infants' Topics
London, Longman
>A pack of materials designed to cater for reception classes (Stage A), middle infants (Stage B) and top infants (Stage C). The materials comprise:

Starter Book A	Starter Book B
Teachers' Book A	Teachers' Book B
Starter Readers	Record Sheets for stages A,
Teachers' Book C	B, C

>Starter books for Stages A and B are in the form of large full-colour posters which are designed for use with groups of children together and are intended to be a basis for discussion leading into the science activities. The materials provide practical suggestions for the teacher and a structure for developing children's skills, language and technological ability.

An Early Start to Science
Richards, R., Collins, M. and Kinkaid, D. (1987)
London, Macdonald Educational
>A resource book offering a collection of science experiences for children aged 5 to 8, to be pursued within the context of integrated infant practice.